# SpringerBriefs in Business

For further volumes:
http://www.springer.com/series/8860

Francesco Schiavone

# Communities of Practice and Vintage Innovation

## A Strategic Reaction to Technological Change

 Springer

Francesco Schiavone
University Parthenope
Naples
Italy

ISSN 2191-5482      ISSN 2191-5490   (electronic)
ISBN 978-3-319-01901-7      ISBN 978-3-319-01902-4   (eBook)
DOI 10.1007/978-3-319-01902-4
Springer Cham Heidelberg New York Dordrecht London

Library of Congress Control Number: 2013947349

Printed on acid-free paper

Springer is part of Springer Science+Business Media (www.springer.com)

*The visionary is the only true realist*

Federico Fellini

# Foreword

An endogenous interest of consumers for old-fashioned things and objects is typical in many industries and markets. Vintage innovation, the subject analyzed in this text, is based on this belief.

If we only look out of the window we see a large number of Fiat 500, Mini, Triumph, or Harley[1] passing; it is easy to understand that in many fields we have a trend to rediscover and propose again lines, brands, and objects with a vintage taste. These products go beyond the present example linked to the car and motorbike sectors; they are interpreted again thanks to modern technologies but they keep and underline vintage peculiarities and habits. If we browse for a moment it is also easy to notice that there are forums and communities for people keen on watches, motorbikes, glasses, hi-fi, musical instruments, and in each commodities' sector category lives in its vintage field a remarkable dynamism and enthusiasm. The involvements of this phenomenon are an opportunity for a company able to find, intercept, (or create) the vintage trends, and able to deal with its structure in the vintage direction. The vintage area is a real competitive arena.

This book summarizes and develops comprehensively the latest and main research theme by Francesco Schiavone, on which the author focused over the past years. The book is organized into five chapters.

Chapter 1 analyzes the phenomena of technological change, adoption, and diffusion of innovation which shape the general context of vintage innovation.

Chapter 2 describes the concept of communities of practices, a typical form of social aggregation in which vintage innovation takes place.

Chapter 3 describes in-depth the phenomenon of vintage innovation, the innovative way by which the so-called vintage products improve the performance of old technology-based products becoming obsolete after technological change. The chapter ends with an analysis of the main implications of this phenomenon for companies in terms of technology management, innovation strategy, and marketing.

Chapter 4 reports two case studies of vintage innovation by companies developing innovations for communities of practice centered around old technology-based

---

[1] For example just to give a clear measure of the phenomenon, during the assembly for the anniversary of 110 years from the birth of Harley, on 13–16 June 2013, the city of Rome was visited by more than 150,000 motorbikes coming from all Europe.

products. The cases here reported are the vinyl emulator for DJs turntablists and film scanners for analog photographers.

Chapter 5 reports two case studies on vintage innovation by communities of old technology users. The cases of multimedia software for radio-amateurs and emulation software for arcade videogamers are presented.

The work done by Francesco Schiavone is an interesting analysis of a growing phenomenon, of a strategic trend supporting companies in their development process and fighting against crisis based on an innovative return to past.

I encourage Francesco to continue his studies and I hope that also other academics will devote to the vintage innovation and to the further analysis of the different social and cognitive implications, strategic but first of all structural, linked to this phenomenon.

Naples, July 2013                                                                    Luca Dezi

# Preface

The word "technology" is derived from the Greek *techne*, meaning art, craft, or skill, and *logos*, meaning word or knowledge. These two meanings well summarize and merge the two basic sides of the concept: the execution of some task or activity and the human narration about this activity.

Technology, as every human activity and every human being, rises, develops, becomes mature, and then declines until the end. And each of these phases shapes some specific narrations and social interactions about and around the practices performed by technology. Many scholars focused, maybe excessively, their attention on the first stages of the technology life cycle and neglected a clear evidence that every day we recognize in television, magazines, or the Internet: people are attracted by the past and old things.

Philosophers as Gianbattista Vico or Alexandre Kojève argue that history repeats itself. Of course, the past shapes the future and somehow, our desires. Economic phenomena as retro-marketing or old product revitalization are based on the need of people living again their past experiences, memories, and emotions. The vintage wave is nowadays mainstream in several markets.

This book identifies communities of practice as the best loci for the celebration of old technology and the resurrection of people's memories about an outdated technology. Everyone of us is member, consciously or unconsciously, of some community of practice with specific legacies, traditions, and memories. When these invisible assets are related to some technological product that became obsolete here comes the phenomenon that I called vintage innovation. Actually, 5 years ago I took the inspiration to develop this research over time, basically, thanks to my former membership of a technological community of practice.

The Merriam-Webster dictionary reports that, literally, the word vintage is an adjective referring to something "of old, recognized, and enduring interest, importance, or quality". Vintage innovation bridges together naturally and innovatively the appeal, nostalgia, and mystery of old technologies with the power and efficiency of new technology. High-tech companies can perform, under given conditions, a technological reverse strategy in order to provide valuable products to their customers belonging to old technology-based communities. The development of a vintage innovation approach is the core of such innovation strategy. However, the empirical analysis reported in Chap. 5 of this book shows vintage innovation

is something deeper and larger than a niche corporate strategy to react to technological change. Indeed, this approach can be implemented even by communities'
members themselves.

This book summarizes the main findings of my research activity of the past
5 years. This was an amazing research experience. Primarily, because the interaction with many members of various communities of practice made data collec-
tion and data analysis interesting and meaningful. They transmitted to me their
enthusiasm and passion for their old beloved products and, as a consequence, this
critically increased my curiosity about this hybrid technological phenomenon integrating old and new technologies, artifacts, knowledge, and competences.

Over this period, I had the opportunity to receive many valuable suggestions
by several colleagues and friends who shared with me the passion for this fascinating topic. I have to thank several (known and unknown) academics across the
world for the development and refining of vintage innovation. First, the European
professors who invited me to give research seminars in their Business Schools:
Ludovic Dibiaggio, Stefano Borzillo, and Renata Kaminska-Labbe at SKEMA
in Sophia Antipolis (2009), Laurent Bibard at ESSEC in Paris (2011), Slawek
Magala at Erasmus University in Rotterdam (2013). I want also to thank all the
editors and anonymous reviewers of the journals in which I published my articles
about this topic (European Journal of Innovation Management, IEEE Transactions
on Engineering Management, International Journal of Innovation Management,
Management Decision, Journal of Organizational Change Management). The
comments of all these colleagues oriented me effectively in framing and defining
the boundaries and implications of vintage innovation.

Furthermore, I wish to express my gratitude to Marina Forlizzi and Maria
Cristina Acocella (Springer Italy), who supported me in the review process of the
proposal of this book and all over the editorial process.

Another special thanks goes to Professor Gaetano Maria Golinelli from
University Sapienza of Rome and Professors Luca Dezi, Adriana Calvelli, and
Franco Calza from University Parthenope of Naples. Also, my colleagues and
friends Michele Simoni, Rocco Agrifoglio, and Jason McVaugh deserve a special
mention. All of them supported me in various moments and steps of my academic
career with great passion and generosity.

Naples, July 2013                                                    Francesco Schiavone

# Contents

**1 Technological Change** .......................................... 1
  1.1 Stylized Facts About Technological Change ................... 1
    1.1.1 R&D Implications ..................................... 4
  1.2 Strategic Reactions to Technological Change ................. 6
    1.2.1 Six-Item Assessment of Technological Competition ....... 9
  1.3 Diffusion and Adoption of New Technology ................... 12
    1.3.1 Factors Impacting on Single Users Adoption ............ 15
  References ...................................................... 21

**2 Communities of Practice** ....................................... 25
  2.1 The Concept of Community of Practice ....................... 25
    2.1.1 Communities of Practice and Knowledge ................ 28
    2.1.2 Communities and Innovation .......................... 31
  2.2 Communities of Practice and Technology .................... 33
    2.2.1 The Social Construction of Technology ................ 33
    2.2.2 The Meanings of Technology for Communities
        of Practice .......................................... 35
    2.2.3 A Technology-Based Typology of Communities .......... 37
  2.3 Reactions of Communities of Practice to Technological Change ... 39
    2.3.1 Resistance .......................................... 40
    2.3.2 Openness ........................................... 43
    2.3.3 Reasons for Ambivalent Reactions ..................... 44
  2.4 Research Agenda About Communities of Practice
    and Technological Change .................................. 45
  References ...................................................... 47

**3 Vintage Innovation** ............................................ 51
  3.1 Introduction .............................................. 51
  3.2 A Third Way to Improve Old Technological Products? .......... 53
    3.2.1 The Saviotti and Metcalfe Model ...................... 54
    3.2.2 Compatibility Between Old and New Technologies ....... 57
  3.3 Vintage Innovation ........................................ 59
    3.3.1 Differences with Similar Approaches ................... 61
    3.3.2 Vintage Innovation and Customers .................... 63

3.4     Vintage Innovation and Corporate Strategies. . . . . . . . . . . . . . . . . .     64
        3.4.1   Technology Management . . . . . . . . . . . . . . . . . . . . . . . . . . .     64
        3.4.2   Innovation Strategy  . . . . . . . . . . . . . . . . . . . . . . . . . . . . .     66
        3.4.3   Marketing . . . . . . . . . . . . . . . . . . . . . . . . . . . . . . . . . . .     67
References . . . . . . . . . . . . . . . . . . . . . . . . . . . . . . . . . . . . . . . . . . .     68

4   Vintage Innovation by Firms . . . . . . . . . . . . . . . . . . . . . . . . . . . . .     71
    4.1     Introduction  . . . . . . . . . . . . . . . . . . . . . . . . . . . . . . . . . . . .     71
    4.2     Vinyl Emulator for Disk Jockeys . . . . . . . . . . . . . . . . . . . . . . . .     73
            4.2.1   Analog Music, Turntables, and Technological Change . . . . .     73
            4.2.2   The Community of DJs Turntablists  . . . . . . . . . . . . . . . . .     75
            4.2.3   Vinyl Emulator . . . . . . . . . . . . . . . . . . . . . . . . . . . . . . .     76
    4.3     Film Scanner for Analog Photographers  . . . . . . . . . . . . . . . . . . .     79
            4.3.1   Analog Photography, Film Cameras, and Technological
                    Change . . . . . . . . . . . . . . . . . . . . . . . . . . . . . . . . . . . .     79
            4.3.2   The Community of Analog Photographers. . . . . . . . . . . . . .     81
            4.3.3   Film Scanners . . . . . . . . . . . . . . . . . . . . . . . . . . . . . . .     82
    4.4     Conclusions and Implications  . . . . . . . . . . . . . . . . . . . . . . . . .     85
References . . . . . . . . . . . . . . . . . . . . . . . . . . . . . . . . . . . . . . . . . . .     87

5   Vintage Innovation by Users  . . . . . . . . . . . . . . . . . . . . . . . . . . . . .     89
    5.1     Introduction  . . . . . . . . . . . . . . . . . . . . . . . . . . . . . . . . . . . .     89
    5.2     Multimedia Software for Radio-Amateurs. . . . . . . . . . . . . . . . . . .     91
            5.2.1   History and Technological Change of Ham Radios . . . . . . . .     91
            5.2.2   The Community of Radio-Amateurs . . . . . . . . . . . . . . . . . .     92
            5.2.3   Multimedia Software  . . . . . . . . . . . . . . . . . . . . . . . . . . .     93
    5.3     Emulators for Arcade Videogames Players  . . . . . . . . . . . . . . . . . .     97
            5.3.1   Arcade Videogames . . . . . . . . . . . . . . . . . . . . . . . . . . . .     97
            5.3.2   Communities of Nostalgic Videogamers . . . . . . . . . . . . . . .     99
            5.3.3   MAME and Other Emulation Software . . . . . . . . . . . . . . . .    101
    5.4     Conclusions and Implications  . . . . . . . . . . . . . . . . . . . . . . . . .    103
References . . . . . . . . . . . . . . . . . . . . . . . . . . . . . . . . . . . . . . . . . . .    105

# Chapter 1
# Technological Change

**Abstract** This chapter summarizes the main stylized facts about technological change. These evidences refer to the main characteristics of the process, the sources of innovation, and the main implications in terms of Research and Development for firms. The main strategic reactions to technological change are presented and discussed: sailing ship effect, shift to new technology, and exit from the market. A multilevel model of analysis is also offered in order to support strategizing and selecting the right option. The chapter ends with a review of the main macro, meso, and micro factors influencing the adoption (or the rejection) of an innovation. The conclusion is that technological, social, and learning factors are often interconnected with each other and draw a complex picture of adoption, diffusion, and substitution. This picture largely affects the impact and speed of technological change.

**Keywords** Technological change • Creative destruction • Sailing ship effect • Old product revitalization • Diffusion of innovation • Resistance to innovation

## 1.1 Stylized Facts About Technological Change

Everyday technology affects and plays some role in our lives. We continuously need and use technology to perform some action or achieve some goal. Technology is "a set of pieces of knowledge, both directly 'practical' (related to concrete problems and devices) and 'theoretical' (but practically applicable although not necessarily already applied), know-how, methods, procedures, experience of success and failures and also, of course, physical devices and equipment" (e.g., electricity or biotechnology) (Dosi 1982, p. 151). Technology refers to the knowledge and/or subcomponents and machineries used by firms in order to develop and assemble their own technological products (or artifacts) for customers. Products differ by technological knowledge as they are commercialized by firms in order to satisfy market needs (Pavitt 1998).

F. Schiavone, *Communities of Practice and Vintage Innovation*,
SpringerBriefs in Business, DOI: 10.1007/978-3-319-01902-4_1,
© The Author(s) 2014

The evolution of technology and technological products is a typical phenomenon in every industry. Technological change is the social and economic process by which an invention becomes a novel technology (innovation) which diffuses within an industry (Schumpeter 1942). Rapid technological change affects and shapes the dynamics of competition and structure of many industries over time.

As early as 1930, researchers had begun to consider the significance of the introduction of new technology to capitalist marketplaces. As already summarised in Schiavone 2011, In that year, the German economic sociologist Werner Sombart (1930) coined the notion of creative destruction, but it was popularized by Joseph Alois Schumpeter who adopted it in his published research. Schumpeter (1942), intellectual father of the evolutionary theory in economics of innovation, was one of the first scholars to recognize how firms contribute to transform an invention into an innovation and to support its diffusion within the market. Schumpeter (1912, 1942) in his seminal studies highlights two main dynamics leading industry technological change: Creative Destruction and Creative Accumulation. Creative destruction (Schumpeter Mark I) is the process through which small firms recombine existing resources in innovative ways in order to launch new products within the market. Entrepreneurs and their firms are, thus, the engines of the development of an economic system. The key process of creative destruction is the recombination of existing tangible and intangible assets in order to develop and commercialize innovations. This process:

> incessantly revolutionizes the economic structure from within, incessantly destroying the old one, incessantly creating a new one. This process of Creative Destruction is the essential fact about capitalism (Schumpeter 1942, p. 83).

The second view (Schumpeter Mark II), instead, suggests that large established firms are the organizations most appropriated into an industry in order to lead economic development as they can invest financial resources in Research and Development (R&D). So they implement a path of creative accumulation, in order to preserve their traditional know-how and technological capabilities.

Technological change is a process involving three different types of actors: firms (offering technology), adopters (individuals or other firms demanding technology), and others parties (e.g., public institutions). All these actors are greatly affected throughout this process by their cultural mindset, expectations, experiences, and external contexts. Consistently with this taxonomy, scholars developed three main theoretical approaches in order to explain the driving forces behind this phenomenon (Dosi 1982):

- The first approach is called technology-push. Authors (as Schumpeter) supporting this view argued that technological change mainly depends on firms (the supply of technological innovation) and their capability to promote innovation into the market. This approach supports, thus, the view of technological determinism (technology drives the development and evolution of society).
- The second view is the demand-pulled approach. According to this view, technological change and diffusion of an innovation mainly depend on market needs and firms' capabilities to satisfy customers through technologically new products (Schmookler 1962).

- A third approach to technological change comes from the evolutionary economics literature (Dosi 1982; Nelson and Winter 1982).[1] This view has its theoretical roots in the Schumpeterian thinking. The scholars of this stream assert innovation as a path-dependent and systemic process in which demand and offer of technology co-evolve. Knowledge and learning affect the interaction of these two parties throughout the process (Rosenberg 1982).

Technology evolves over time within specific technological trajectories and depends on the organizational routines of firms and on their continuous search for new technological solutions to offer to the market. Evolutionary economics scholars were the first to explain why old technologies survive for some time after a new technology starts diffusing into a market (Nelson and Winter 1982; Dosi 1982; Rosenberg 1982). Nelson and Winter (1982) developed a two-technology evolutionary model in order to describe the process by which a new technology replaces an old one at country level. Three different forces can accelerate or delay substitution within an economic system: production cost differentials, expectations patterns of entrepreneurs, and availability of technological complementarities and infrastructures (Rosenberg 1994).

Dosi (1982) develops three critical concepts to explain how technological change occurs over time: technological paradigm, technological trajectories, and technical progress. A technological paradigm is a "model and a pattern of solution of selected technological problems based on selected principles derived from natural sciences and on selected material technologies" (Dosi 1982, p. 152). This concept is very close to the notions of technological regime (Nelson and Winter 1982) and technological guidepost (Sahal 1985). Every technological paradigm has a great exclusion effect as it makes "blind" engineers and designers toward other technological solutions not comprised within the paradigm. Dosi assumes a technological paradigm as a sort of research program that provides and suggests to scientists some research lines to follow instead of others. When an incumbent technological paradigm change occurs a techno-paradigm shift, which refers to radical changes in the way technology has been and continues to be developed, is applied and commercialized over time (Kodama 1995). In the field of technology, Dosi considers these lines as technological trajectories, "the pattern of normal problem solving activity on the ground of a technological paradigm" (Dosi 1982, p. 152). A technological trajectory is drawn by "a cluster of possible technological directions whose outer boundaries are defined by the nature of the paradigm itself" (Dosi 1982, p. 154). Within the same paradigm, technological trajectories have usually strong complementarities for each other. The last critical concept stressed by Dosi is technical progress, which is the improvement over time of the trade-offs among the technological and economic variables which the paradigm defines as relevant. A typical trade-off is the one between price and quality of goods.

A basic example to understand the meaning of these concepts comes from the music industry. The analog sound and digital sound may be considered as two different technological paradigms satisfying the same problem (or users need) within the

---

[1] "The core concern of evolutionary theory is with the dynamic process by which firm behaviour patterns and market outcomes are jointly determined over time" (Nelson and Winter 1982, p. 18).

music industry. Over time, more technological trajectories emerged within each paradigm: vinyl and audio-cassette in the analogical paradigm; compact disk and mp3 in the digital paradigm. In this case, technological progress over time may be referred, for instance, to the trade-off between the quality of sound and the price of audio supports.

A number of studies on dominant design analyzed the subject of technological change (e.g., Abernathy and Utterback 1978; Anderson and Tushman 1990; Murmann and Frenken 2006). The shift toward a new technological paradigm starts when a technological discontinuity emerges (Anderson and Tushman 1990) and new technological trajectories, technically more advanced and based on different scientific notions, stem from it. Technological discontinuity is based on any breakthrough[2] that advances an industry price versus performance frontier (in other words, the industry technical progress) (Anderson and Tushman 1990, p. 604), and shapes a new technological cycle. Every cycle starts with a ferment era in which emerging technological formats compete in order to become the new dominant design. This is a product widely adopted within the corresponding industry and its emergence apparently changes the nature of the market competition (Abernathy and Utterback 1978).

A dominant design changes the characteristics of innovation and competition within an industry as it establishes new architecture specifications within its product category. The predominance of one technology over its competitors starts an era of incremental change over which firms focus on the improvement of the dominant technology's performance. This phase lasts until a new breakthrough occurs and a new technology, competing with and winning against the incumbent one, emerges. Afterward, a new technological cycle with a new ferment era restarts and a new "standard war" (Shapiro and Varian 1999) for the identification of its new dominant design is launched.

Technological cycles evolve as an S-curve over time. A common view between scholars is that the new competing technology emerges when an incumbent one approaches its "performance limits" and is not able to experience any further improvement. The S-curve model of Foster (1986) shows how the level of performance of a technology changes over time and how technology competition occurs (Fig. 1.1). After a period of co-existence between an old technology (To) and a new one (Tn), the latter starts offering superior performances at a given moment t1. After some time, Tn replaces To totally (which has arrived to its performance limit) in the market and becomes its dominant technology. The process of technological change and the technological cycle of To are completed.

## 1.1.1  R&D Implications

The S-curve model by Foster is based on the assumption that competition is the unique possible relationship between old and new technologies. However, the nature of this relationship depends on the specific choices and behaviors of both

---

[2] Breakthroughs are revolutionary innovations sustaining the growth of the market. Henderson and Clark (1990) distinguish four different types of product innovations: radical, incremental, modular, and architectural.

**Fig. 1.1** Schema of
technology competition
(adaptation from Foster 1986)

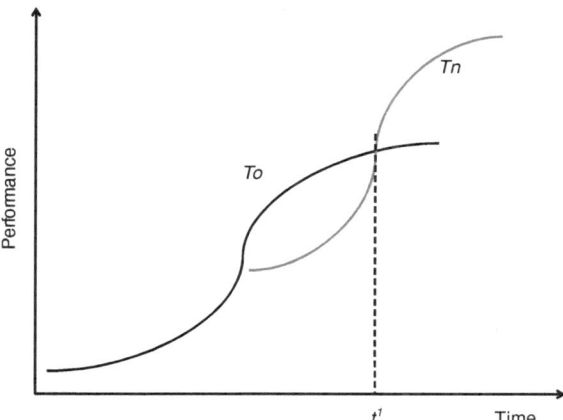

users (organizations and individuals) evaluating to adopt or reject new technology
(the demand of technology) and firms engaged in R&D and commercializing technological innovations (the supply of technology).

Referring to the latter, a rich body of literature in the field of R&D management
(e.g., Kodama 1995) on the topic shows that the nature of relationships between
old technology and new technology may be not just competitive but also collaborative. The evolutionary perspective by Patel and Pavitt (1994) suggests that
"new paradigms do not destroy old ones but complete and extend them." The history of old technology is critical to understand present technologies and forecast
their future developments (Edgerton 2006). The acceptance of this view implicitly
involves every new technological product as a sort of *compromise* between some
former technologies and some future scientific and/or technological advancements
or some new human needs and problems to solve (Saviotti and Metcafe 1984).

A variety of hybrid behaviors, affecting the likelihood of achieving a techno-
paradigm shift, between competition and collaboration has been noticed in the
R&D approaches of organizations. There is competition if companies simply
develop new technology achieving superior performance and replacing the old one
in the market. Companies usually establish competition between technologies by
developing technological breakthroughs. There is collaboration if firms developing
new technology integrate it with the old technology. Within one firm, the extent of
R&D interdependency and interaction between an old and a new technological
paradigm can be of three types (Freddi 2009):[3]

- new technologies develop independently and autonomously by old ones in the
  same company or sector;
- there is technological complementarity between old and new technological paradigms, but technologies are still independent;
- firms opt for technology fusion between the old and new technological paradigms.

---

[3] However, the author reports that "the boundaries between the three cases are rather fuzzy and a
clear distinction is only possible in theory" (Freddi 2009, p. 550).

**Table 1.1** Different R&D approaches (Author's elaboration on Kodama 1995)

| R&D approach | Evolutionary logic | Type of innovation sought | Techno-paradigm shift |
| --- | --- | --- | --- |
| Technology fusion | Technological continuity in more (previously separated) industries | Incremental innovation | Not probable (at least in each single industry) |
| Breakthrough | Technological discontinuity in one industry | Radical innovation | Very probable |

Technology fusion is a critical R&D approach for establishing collaborative relationships between different technologies. This is an approach "combining existing technologies into hybrid technologies" (Kodama 1995, p. 200). It is a different notion from complementarity as it implies that "the reciprocal dependency of the different technologies is so high that they are actually fused together in one (new) body of knowledge" (Freddi 2009, p. 550).[4]

The success of technology fusion depends on the capabilities of technology integration held by the firm's management and R&D department. Technology integration is "the set of investigation, evaluation and refinement activities aimed at creating a match between technological options and application context" (Iansiti 1998, p. 21). Indeed, it is not possible to implement technology fusion if there is no synergy between different kinds of engineering sub-disciplines. A typical example of technology fusion is mechatronics that combines mechanical technologies with electronic and material technologies.

Technology fusion is an alternative R&D approach to breakthroughs. Summarizing, firms planning their R&D strategy should opt for a technology fusion (collaboration between technologies) or breakthrough approach (competition between technologies). This choice will affect the type of innovations firms will develop, and the likelihood of reaching a more advanced techno-paradigm shift. Table 1.1 reports some basic characteristics of each approach.

## 1.2 Strategic Reactions to Technological Change[5]

Technological change and the substitution of old products with new ones are common events in every industry, but events that may have significant strategic and competitive implications for firms. The literature acknowledges that an improper

---

[4] In this sense, Kodama (1995, p. 203,) argues that "in fusion, one technology is added to another to come up with a solution greater than the sum of its parts. In other words, fusion implements an arithmetic in which one plus one makes three. Fusion is more than complementarities because it creates new markets and new growth opportunities for each participant in the innovation."

[5] This section is based on and extends some elements of the analysis developed by the author in Schiavone (2011)

management of the reaction to technological change might bring about even the failure of a market leader firm. For instance, 13 of the 17 firms (except IBM, Fuji, Hitachi, and NEC) operating in the hard disk industry prior to 1975 failed or were acquired within the following 20  years (1976–1995). This industry has been characterized by continuous innovation of the products architecture since its birth (mid-1950s) (Christensen 1997). The decline of so many established companies was not an outcome of the pace and complexity of technological change (the so-called "technology mudslide hypothesis") but the result of their excessive focalization on customers' needs and partial underestimation of the architectural technological innovations offered by new entrants (more than one hundred in those 20 years). The critical advantage of their disk drives was the physical size, much smaller than existing products. Most of the incumbent firms did not perceive properly this change, did not follow the new technological trajectories emerging into the industry, and thus, declined over time.

The case just outlined shows that the defining features of success in technological competition depend on the ability to promptly and effectively react to technological change. Analysis of the strategic alternatives incumbent firms can implement when technological change occurs (Cooper and Smith 1992; Howells 2002; Adner and Snow 2010) is a popular topic in the existing literature on technology competition.

While the theory of Creative Destruction provides a useful overview of the environment in which technological competition takes place, Adner and Snow (2010) highlight two potential macro-level strategic reactions for incumbent companies for the diffusion of a new technology: racing strategies and retreat strategies.

The first category are racing strategies by which firms try to increase the performance of their older technology in order to reduce the disparity created by diffusion of new technologies. Racing strategies usually imply the implementation of old product revitalization, an R&D technique "used to give a new lease of life to an existing product by bringing it up to date in its design (styling), performance, costs, or features" (Gaskell 1992).

The second category of macro-level strategic reactions are retreat strategies, by which firms accommodate the entry of new technology, primarily repositioning (and to a minor extent revitalizing too) their traditional products based on the old technology. The repositioning occurs by defending the position of the old product or relocating the product to a new market.

Howells (2002) proposes three main strategic alternatives when incumbent firms, manufacturing the old technology, face technological change:

1. exit from the old market;
2. switch toward new technology;
3. the sailing ship effect.

Exit from the old market is the most drastic reaction possible for technological change. In some cases, exit is the most appropriate reaction following drastic market shrinkage. Otherwise, this strategic reaction may reflect the disinterest (or incapability) of the firm to face competition under new market conditions and technological standards. The adoption of this reaction implies that the firm needs to search for one or more new markets in which to enter and re-invest the resources dismissed from the old one. In their study on the effects of technological change on the structure of the American tire industry, Klepper and Simons (2000) found that small and young firms tend to exit from the market when a new technology emerges. So while in some cases it may be a small firm that leads technological innovation in a market, Suarez and Utterback (1995) note that firms entering into an industry many years before the emergence of the dominant design are likely to face and survive technological change better than younger organizations. Suarez and Utterback explain this evidence by arguing that older firms have more resources to experiment with during periods of fast change.

The switch toward new technology is a strategic reaction more frequently followed than exit. In this case, the firm renews its product portfolio by developing and commercializing new products based on the emerging technology and its paradigm. In this way, the firm contributes to the *Creative Destruction* of the former market equilibriums. The switch is a complex process based on several decisions concerning corporate strategy, the organizational structure of the firm, and the psychology and perceptions of its managers. Even with superior resources, large and established firms may fail in the implementation of this strategy (Utterback 1994). A well-known case of an unsuccessful switch toward a new technological paradigm is Polaroid. This American company was not able to follow the technological change that occurred in the camera industry during the 1990s (the transition from instantaneous/film-based pictures to digital ones). The main reason behind this failure was the poor cognition of Polaroid management in both understanding the ongoing technological evolution in the industry and trying to set up an adequate search for new technological solutions to face that radical change (Tripsas and Gavetti 2000).

The sailing ship effect is the "acceleration of innovation in the old technology in response to the threat from the new"[6] (Howells 2002, p. 887). This option is an economic version of what biologists define as the "red queen effect" (Barnett and Hansen 1996). This effect occurs when firms attempt to preserve their own technological competencies from a decline due to technological change. Its name derives from what happened in the naval industry in the second half of the 1800s: an increase in the innovation and performance of sailing ships occurred after the introduction of steam ships. The explanation of Gilfillan (1935) was that sailing ships producers perceived steam ships as a threat by which their competitors were able to surpass them. Such firms responded by revitalizing and repositioning their older technology within the broader product market. The product is no longer commercialized in the mass market, but instead is positioned as a niche product for those groups of users not interested in adopting the new technology (Schiavone 2013). The sailing ship effect, thus, occurs when an old technology is revitalized

---

[6] Another definition of this phenomenon reports that sailing ship effect is "process whereby the advent of a new technology engenders a response aimed at improving the incumbent technology" (De Liso and Filatrella 2008, p. 593).

and experiences a "last gasp" due to the risk of replacement by a new substituting one. Three possible explanations of this effect are provided by Snow (2008):

- Try Harder: old technologies are improved in order to survive and to avoid being replaced by new ones. This is the most straightforward explanation for the sailing ship effect.
- Spill over: incumbent technologies' efficiency can improve even without technological change and replacement. Components of the new technology "spill over" in products based on the incumbent technology.
- Selection: the substituting technology generates notoriety for the old technology from novel applications rather than from technological innovations.

The categorization of Howells (2002) is based on the study of industry technological change and corporate experiences. However, its categories are based on the assumption that the company is just interested in one market (the old or the new one) or in none of them (in threports that sailing ship effect ise exit reaction). Many cases indicate that after a new substitute technology emerges firms may continue to produce and sell their old products for a long time, while also trying to develop and commercialize new technology products (such as by producing and selling both analog cameras and digital cameras). Howell's taxonomy does not include this possibility, a fourth type of strategic reaction: a combination of sailing ship effect and switch (Schiavone 2011). In this way, companies both keep their presence in the old market and enter in the new market.

Cooper and Smith (1992) recognized a similar behavior by analyzing the reactions of 27 market leading companies threatened with technological substitution. Most of these companies reacted to the emergence of radical innovations by developing new products while continuing to produce and commercialize the older ones for a long time. Cooper and Smith (1992) define this reaction as participation strategy. Dominant firms in old markets attempting to implement this strategy must decide: (1) the time of entry in the new business; (2) the extent of their commitment; (3) the competitive strategy, and (4) the degree of organizational separation between old and new operations.

## 1.2.1  Six-Item Assessment of Technological Competition

To reach a high-quality decision on strategic direction in response to the introduction of a new technology, it is important that managers systematically assess their internal and external competitive positions. Reviewing the existing literature on technological competition, it is possible to simplify this assessment by considering the organization's stance with regard to six key items:

1. *Possibility of old product revitalization*: This R&D approach gives an interesting possibility to managers for the exploitation of existing assets and resources. If the products based on the old technology are hard to develop further, then incumbent firms will find it difficult to keep operating and innovating in their

traditional market. Instead, if technical developments of the old products are possible, managers could opt for the implementation of revitalization or retro-marketing strategy in order to propose technologically upgraded versions of these products to nostalgic niches of users.

2. *Extent of disruption of innovation*: This condition affects the strategic reaction to change. Innovation can disrupt and destroy the existing paradigm of an industry. Abernathy and Clark (1985) suggest two main domains of the innovative activity that are affected by the type of emerging innovation: technology and market. If a disruptive novelty changes both the market/customers linkages and technological competencies of an industry, then many incumbent firms may find it harder to switch into the new market (Christensen 1997). In other cases, innovations will produce significant changes in just one domain (market/linkages or technological competencies) and the extent of their disruption will be minor. The extent of disruption will be high if innovation changes both market relations and technological competencies in the industry, or low, if just one level of change occurs.

3. *Old market competitive position*: The legacy of a firm's competitive position in the old market affects what the company will do in the future. Leading firms in the old market are likely to benefit even after technological competition is introduced. A dominant market position should be positively associated, for instance, with brand awareness, customers' loyalty, corporate reputation, and/ or the extent of distribution channels. In this case, old market leaders could still continue serving some of their loyal customers who resist innovation (MacVaugh and Schiavone 2010). Commercial and/or scientific partnerships could be still exploitable over time if new products did not destroy the old market's value networks. On the contrary, if the firm held a follower position in the old market and cannot exploit any further competitive advantage from such market conditions, then managers should focus corporate resources on the emerging market. The disruptive emergence of a new technology reshapes (and often decreases) the old market and creates a new one. The potentially disruptive power of new technology indicates the need for a precise (and comparative) reassessment of the competitive environment.

4. *Difficulty in entering in the new market*: Transition to the new market is not an automatic step for an incumbent organization. The extent of the difficulty in entering a new market is primarily an issue of entry barriers (e.g., political constrains or investments) that firms have to afford and overcome. Different firms are likely to experience different levels of difficulty in performing the transition. For instance, the lack of commercial licenses, patents, or the inaccessibility of strategic information may increase the difficulty of entering the new market, and thus, make the transition more risky.

5. *Dynamic capabilities*: Organizational capability is the "ability to perform repeatedly a productive task which relates either directly or indirectly to a firm's capacity for creating value through effecting the transformation of inputs into outputs" (Grant 1996). Firms willing to follow technological evolution should change their internal processes and organizational routines implemented

over the former technological paradigm. To this end, they must be able to interpret the emerging technology evolution and learn its related knowledge. Dynamic capabilities are "the firm's ability to integrate, build, and reconfigure internal and external competences to address rapidly changing environments" (Teece et al. 1997). Dynamic capability is also directly related to the level of prior knowledge. If organizations hold dynamic capabilities, then the exploitation of the opportunities related to the new technology and the survival in the new market environment are easier.

6. *Old core capabilities*: The value of capabilities used in the old market is another crucial factor to consider. A valuable type of capability for a company, in terms of competition, is a "core" capability. This notion refers to "the knowledge set that distinguishes and provides a competitive advantage" to its holder-company (Leonard-Barton 1992). The same author states that four dimensions shape core capabilities: skills and knowledge base, technical systems, managerial systems, and values and norms. If the company held any core capability in the declining market, then it could still partly try to exploit this knowledge set in the future in order to continue implementing those activities based on that bundle of knowledge. Core capabilities often might be a sort of "core rigidities" as they may generate an "incumbent inertia" to change in companies and make harder their switch to the new technological paradigm.

The six-item analysis method includes three variables linked to a company's history (old core capabilities, old technology revitalization, and competitive position in the old market) and the other three refer to its emerging context (dynamic capabilities, easiness of new market entry, and extent of new technology disruption). Table 1.2 indicates how these six variables form the basis of a very simple survey, which could be administered in times of technological competition. Managers should reply to a multiple-item Likert scale by assigning a score between 1 and 5 to each statement according to their agreement or disagreement with the sentence. Each item measures corporate strength against a specific variable.

**Table 1.2** A multiple-item scale to support decision-makers

| Statement (Likert scale) | Score I disagree/I agree |
| --- | --- |
| 1. My company produces a product which still has significant development potential | 1–5 |
| 2. Recent innovation will not disrupt the linkages of my company with its traditional customers or our set of industrial competencies | 1–5 |
| 3. My company was a leading and well-recognized organization in the old technology market | 1–5 |
| 4. My company is not going to experience any particular issue in entering in the new technology market | 1–5 |
| 5. The core capabilities of my company have been critical in order to achieve a competitive advantage in the old technology market | 1–5 |
| 6. My company is able to change its capabilities in a dynamic way, following the course of technology evolution | 1–5 |

**Table 1.3**  Multiple-items total score and preferable strategic reactions

| Total score | Advisable strategic reaction | Complexity of implementation |
| --- | --- | --- |
| Less than 14 (or 14) | Exit | Low |
| Between 15 and 22 | Sailing ship effect | Medium |
| Between 15 and 22 | Switch | Medium |
| More than 23 (or 23) | Sailing ship effect + switch | High |

As the scale is composed of just six statements (thus, one statement for each variable) and the range of each score is between 1 and 5, then the minimum potential score is 6 and the maximum is 30. The range between these two extreme scores is 24 points.

Depending on the context and the disruptive potential of the technological change, if the total score is between 6 and 14, then the company is weak and its chances of success in managing and effectively facing technology competition are probably few. If the total score is between 15 and 22, then the firm is basically 'in the middle' which means it will survive in the market if able to understand and implement the changes others are making in response to new technology. If the final score is 23 or higher, the company is likely one of the shapers of the marketplace, and thus, can afford a more complex reaction. Of course, the specific characteristics of the firm, technology, and industry should inform managers' construct of a more context-specific analysis. The final score of the multiple-item scale is also useful in orienting managers toward a strategic reaction (Table 1.3).

Exit is advisable for incumbent firms with a score of 14 or lower, whereas switch and/or sailing ship effect are likely to be good suggestions for companies achieving a score between 15 and 22. The selection between these two reactions will be based upon the evaluation of the scores achieved in each item. For instance, sailing ship effect should be preferred over switch if the company achieved better scores in the items related to the old market than in those related to the new market. Finally, the combination of sailing ship effect and switch is the reaction most difficult to implement, and so considering the complexity involved, only market leaders (with a total score higher than 22) should take this path.

## 1.3 Diffusion and Adoption of New Technology[7]

The main limit of many industry-based studies on technological change is that they, which focus their attention mainly on firms, do not elaborate what motivations really matter in order to explain why some new technologies diffuse in a market and are adopted by users, while others are not. For instance, the common response of evolutionary economics to this point simply focuses on the extent of solving problems (or satisfying users' needs) of new technologies.

---

[7] This section is based on and extends some elements of the analysis developed by the author and his co-worker in MacVaugh and Schiavone (2010)

The theory on diffusion and adoption of innovations provides useful insights into customers' decisions when technological change occurs. After firms embody their knowledge and capabilities into a new technology, their main concern is the speed at which it diffuses into the market.

Diffusion of an innovation is the last step in technological change. The speed of diffusion greatly depends on how much time is needed for the emergence of a dominant design (Anderson and Tushman 1990). The literature on diffusion of innovations is rich (e.g., Rogers 1995; Grubler 1997; Geroski 2000; Hall 2004). Rogers (1995) defines the diffusion of an innovation as "a process by which an innovation is communicated through certain channels over time among the members of a social system." Adoption is "a decision to make full use of an innovation as the best course of action available" (Rogers 1995). Diffusion, thus, is given by the sum of more single user adoptions within a market.[8] It is a spatially and temporally bounded phenomenon depending on the choices of both individual and organizational actors to adopt an innovation at one point in time. Diffusion is a slow but dynamic process, often bringing about outcomes hard to predict precisely, and follows an "S-shaped" growth path. In this light, Rosenberg (1972) argues that:

> In the history of diffusion of many innovations, one cannot help being struck by two characteristics of the diffusion process: its apparent overall slowness on the one hand, and the wide variations in the rates of acceptance of different inventions, on the other (Rosenberg 1972).

Diffusion and single adoptions of innovations do not occur within all markets simultaneously. Scholars distinguish various ideal types of adopters according to their rapidity in becoming users of the innovation compared to the rest of the market (Rogers 1995)[9]: innovators, early adopters, early majority, late majority and, finally, "laggards" (the latest individuals to adopt the innovation). According to this theoretical viewpoint, adoption is part of a five-stage process: (1) awareness of the innovation; (2) interest; (3) evaluation; (4) trial; and (5) user adoption. Therefore, diffusion of an innovation is not successful if the new technology is not interesting for or is rejected by most of its potential users.

Diffusion and adoption of innovation are very often linked to the substitution of an incumbent technology. Various epidemic models were proposed in marketing sciences in order to consider both diffusion of innovations and technology substitution over time (e.g., Bass 1969; Norton and Bass 1987; Geroski 2000). Most of these studies analyzed substitution in terms of changes in technology sales over time. An old technology can continue its market penetration

---

[8] On the difference between diffusion and adoption Majumdar and Venkataraman (1993, p. 522) write that these "are different dimensions with which to analyse the same underlying phenomenon. What distinguishes one from the other is the unit of analysis, the level of aggregation and the time frame over which each dimension is analysed."

[9] Of course, Rogers reports also that within a market there will be even some individuals not interested in becoming adopters of the innovation.

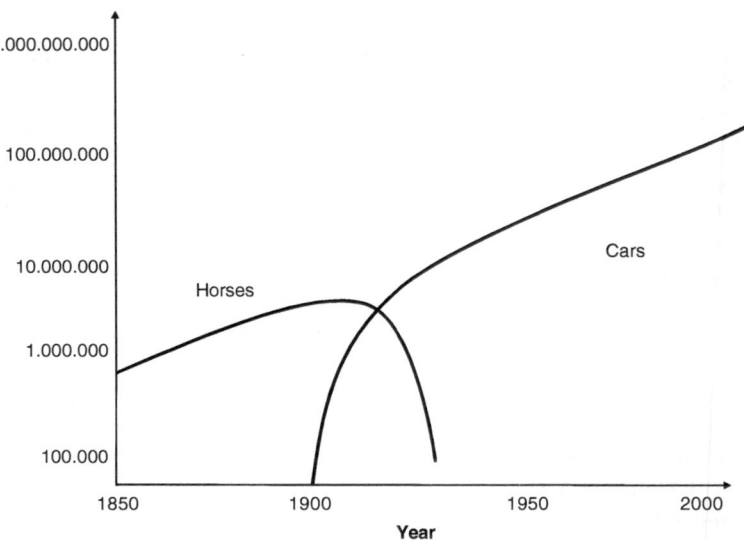

**Fig. 1.2** Number of non-farm draft animals and automobiles (adapted from Grubler 1997)

for a long time although a new product of a further technological generation has already started to substitute it in terms of sales (Norton and Bass 1987). Old technologies survive in the market for a long time even after new technologies have been launched. An economic explanation of this resistance over time of old technologies is given by the fact that new and old capital are complementary inputs and people invest in old technologies even when new ones are available (Chari and Hopenhayn 1991).

Other contributions attempted to define the better entry-time decisions for new technologies substituting old ones. For instance, in the case of monopolistic industries, it has been proved the "now or at maturity" rule (to launch the innovation as soon it is available or when the old technology market is mature) is the optimal timing strategy for the monopolist (Mahajan and Muller 1996).

Probably within the recent history, the best example of technology substitution is the slow but unstoppable process by which utilization of automobiles replaced utilization of horses for private transportation (Fig. 1.2) (Grubler 1997). Adoption, diffusion, and substitution are interconnected processes that might bring about very different final outcomes. For instance, some users may decide for some reasons to be laggard in the adoption of innovations based on new technological paradigms. Others may adopt innovation without substituting old technology. These users will utilize both (old and new) technologies and their products in order to perform a unique activity in their daily life. A banal example of this behavior is given by the simultaneous integration by people of more technologies based on different and substitutive technological paradigms (e.g., post mail, e-mail, fax, telephone, video chat) to satisfy personal communication needs.

### 1.3.1  Factors Impacting on Single Users Adoption

Many scholars, focusing their analyses just on individuals' adoption (rather than on organizational adoption) of new technologies, elaborated an amount of theoretical models aimed at understanding what conditions affect people behaviors, decisions, and perceptions (e.g., Davis 1989; Bagozzi et al. 1992; Venkatesh and Davis 2000; Venkatesh et al. 2003). For instance, a well-known framework is the "Technology Acceptance Model" (TAM), in which authors propose two types of critical determinants of users' behavioral intention (Davis 1989):

1. *Perceived usefulness*: The degree to which a person believes that using a particular system would enhance his or her job performance;
2. *Perceived ease of use*: The degree to which a person believes that using a particular system would be free of effort.

The key condition for new technology acceptance is its capability to satisfy effectively users' needs with no relevant efforts. If these components are not satisfied by innovation, then there are no acceptance, new technology adoption, and old technology substitution.

Ram (1989) argues that behavioral resistance to innovations is due to users' perceived risk about innovation and their habits. Perceived risk is the risk that a consumer perceives in adopting an innovation. It is classified into four components: psychological, economic, social, and functional risk. Habit (or cognitive risk) is given by (1) the amount of cognitive processing that a consumer needs to do in the context of innovation and (2) consumer beliefs about innovation (if it is against his/her values or norms then resistance occurs). If these conditions are positive for innovation, then its adoption should not be problematic.

An individual may be a fast or laggard adopter for several reasons. A number of studies highlight that diffusion of innovations within a system and single users adoptions are influenced by both technological-industrial and socio-cultural factors (Rogers 1995; Anderson and Tushman 1990; Grubler 1997; Rosenberg 1982). Grubler (1997) argues that diffusion depends upon social, economic, and technical factors. If they jointly support diffusion over time, users will substitute the old technology with a newer one, reducing gradually the utilization of the former in the market. Another perspective reports that the actor's decision to adopt an innovation depends on the characteristics of the innovation itself, innovators, and the external environment in which diffusion takes place (Wejnert 2002).

In sum, the literature outlines three levels of factors (macro, meso, and micro) affecting this single user decision. This interpretation does not aim at providing a general theory of technology adoption but just presents a (not exhaustive) list of the main factors affecting, at various levels, the people's decision of adopting new technology or continuing still to use an older technology.

The rationale of this framework is the more new technology befits users' needs and goals according to these three types of factors, the more its strength versus former technologies is great, and the more is the likelihood that a competitive relationship of substitution will take place between them. Conversely, if new technology fails in satisfying technologically, socially, and cognitively its potential users, then technological change should be slower and both the supply and demand sides may find it valuable to establish some kind of collaborative relationship between new technology and old technology.

### 1.3.1.1  Macro Factors

Macro factors refer to industrial characteristics (e.g., industry size or technological infrastructures) that affect in some way single user adoption of new technology. Technological factors help in understanding at "macro" level new technology adoptions. For instance, *network externalities* are a typical industry characteristic that significantly affect the adoption and diffusion of an innovation, respectively, at both individual and market levels. This construct refers to the utility an individual adopter of an innovation achieves by the increase in the total number of adopters in the technological market (Shapiro and Varian 1999). Classic examples of how network externalities facilitate the diffusion of a new technology include the Telephone, Fax, or Internet. Similarly, technological conditions are critical in the so-called network industries[10] (e.g., broadcasting, video games, and air transports), in which the adoption of an innovation is not possible if specific technological infrastructures (e.g., televisions, personal computers, or airports) are available and work (Shy 2001).

New technological products entering into a market rarely are completely new or hold a "stand-alone" value. More often their design is shaped by other substitutor and/or complementor technologies or market products[11] (Brandenburger and Nalebuff 1995). The availability of *complementary technologies* and assets positively affects the adoption of new technology (Teece 1986). For instance, the rate of adoption of USB pen-drives (technological device substituting hard-diskettes) was strongly dependent on the prior diffusion of USB ports in the personal computer market. Similarly, the diffusion of operating systems has historically been correlated with the amount of software available for it to run. This explains why recent releases of various open sources operating systems (e.g., Ubuntu) comprise compatible free software packages as well.

The problem of technological complementarity is related to the presence of an *industry standard* (dominant design), another critical factor affecting the adoption

---

[10] A network industry is a market with the following characteristics: complementarity, compatibility, and standards; consumption externalities; switching costs and lock-in; significant economies of scales in production (Shy 2001).

[11] "Substitutors are alternative players from whom customers may purchase products or to whom suppliers may sell their resources... complementors are players from whom customers complementary products or to whom suppliers sell complementary resources." (Brandenburger and Nalebuff 1995, p. 60).

of radical technological innovations. Indeed, dominant design links together a network of various complementary technologies (e.g., CD players, compact discs, optical disk readers and software). This makes it harder for users of an existing technology (and its technological network) to adopt a newer and completely different product satisfying the same needs, as the utilization of the innovation requires complementary technologies not yet widespread in the market.

Therefore, the technological complexity of an existing dominant design reduces its retire-ability as well. If a complex product is an artifact bridging together more levels of technologies (each one with specific design settings) (Murmann and Frenken 2006), then users may find it both risky and/or problematic to shift to a different technology made by technological subsystems utilizing different components not yet spread into the market. This brings about certain user expectations of the success of an emerging technology; thus its network of complementary products will affect its rate of adoption (Gandal et al. 2002).

Finally, even regulations and policies may hamper or sustain the entry of a new technology and the replacement of an old one within the market. For instance, the adoption of digital terrestrial television platforms spread widely in many European countries solely after the reforms of their national radio television systems.

If macro factors do not induce individuals to change, they should prefer to use old technology and delay to adopt new technology.


### 1.3.1.2  Meso Factors

A second level of factors affecting adoption stems from the social system in which the potential adopters are embedded (Grubler 1997; Geels and Schot 2007). The assumption that this level of factors affects choices of technology adoption by users entails the opposition to the idea of technological determinism promoted by some scholars in the past. Meso factors refer to the features and internal dynamics of social groups (e.g., values or orientations) to which potential users belong and affect in some way the predisposition and quickness of individuals to adopt new technology or products. Social groups are one of the key dimensions of the multilevel perspective of technological transition (Geels 2005). The main meso factors affecting new technology adoption are: group cultural and relational specificities, processes of social construction of the technology, group openness toward innovation, opinion leaders' orientation about new technology, and social contagion within the group.

Cultural and symbolic specificities (e.g., norms, values, and traditions) widely shared within the groups to which users belong are critical for the individual adoption of new technology. For instance, in the 1900s the new cultural landscape within the United States (very enthusiastic toward the opportunities of social progress offered by electricity) facilitated and speeded up the transition from horse-drawn carriages to electric-engine vehicles (Geels 2005). Adoption, diffusion, and technological development is affected by a process of social construction of technology between the main social groups of a socio-economic system (Olsen and

Engen 2007). In this view, every new technology is the final result of a process of negotiation between these social groups representing different interests and problem definitions. This is the key idea of the social construction of technology (SCOT) approach (Pinch and Bijker 1984). If a new technology does not fit the interests or values of a given social group, their members will be likely to avoid its adoption. The adoption of an innovation thus depends on an individual sense-making process that any potential adopter undertakes every time he recognizes that the artifact may satisfy his needs and be socially accepted by his social system.

A group culture open toward innovation affects individual adoption of new technology. Niches of users are critical for adoption and development of innovations. Von Hippel (1988) coined the notion of lead users and found that these individuals form a particular niche of product users able to perceive, thanks to their professional activity, future needs of the mass market before they are generalized and perceived by others. For instance, this was the case of CAD designers that greatly supported companies with their suggestions in order to develop new successful circuit boards for the rest of the market. Recently, also Geels and Schot (2007, p. 401) stressed the criticality of niches for technological change by introducing the concept of technological niches, "small networks of actors [that] support [radical] novelties on the basis of expectations and visions" and act as "incubation room" by protecting novelties against mainstream market selection.

Relational specificities of the group (e.g., hierarchies) are another important meso factor. For instance, the status achievable within a social system by an individual through the use of an old technology or product influences his propensity to adopt newer products (e.g., expensive analogical watches often are status-symbols much more exclusive than digital watches). The risk of a new technology rejection probably might be lower if the first adopters are the group opinion leaders (or the champions in case of an organization). These individuals play a critical role in the diffusion of innovations (Rogers 1995).[12] If the use of a technological innovation is negatively accepted or misunderstood within a community, the rate of its adoption is likely to be slowed too. Firms usually ask for opinion leaders' support in order to prevent this risk. A positive appreciation of opinion leaders (if they exist) is critical for expanding the social acceptance of technological innovation within their community (or market). However, adoption of innovation can also be considered as the outcome of a "bandwagon" process relying on the reciprocal contagion between the "peer" nodes of a social group. Social contagion is the process by which a person catches an idea or behavior from another person (Burt and Janicik 1996). It is a specific feature of networks and is commonly operationalized through cohesion and structural equivalence, two typical network measures considered as the driving mechanisms of contagion. Medical innovation is a well-known example of how social contagion can determine the dynamics of adoption of a technological innovation within a community of users (Coleman, Katz and Menzel 1966). Contagion and personal preferences of doctors are equally critical to orient adoption of innovations in medical communities (Burt 1987).

---

[12] Opinion leaders are individuals who frequently influence others' attitudes and behaviors in adopting an innovation (Rogers 1995).

If meso factors do not induce individuals to change, they should prefer to use old technology and delay the adoption of new technology.

### 1.3.1.3  Micro Factors

These two types of factors outlining adoption of new technology is a multidimensional process resting on the evaluation of both *hard* and *soft* conditions. However, a third level of analysis should be considered. Micro factors refer to the cognitive, psychological, and personal characteristics of individual users (e.g., ability to learn new technological competencies) that influence in some way their adoption of new technology. Knowledge and perceptions of users are the main two dimensions affecting this third type of factors and both are related to the notion of switching costs. They refer to "search costs, transaction costs, learning costs, loyalty customer discounts, customer habit, emotional cost and cognitive effort, coupled with financial, social and psychological risk on the part of the buyer" (Fornell 1992). Shapiro and Varian (1999) in their well-known book on information economics stress how the extent of knowledge and capabilities of individuals using an existing technology can hamper their adoption of a new technology aimed at substituting the former one due to a lock-in constraint.[13] Therefore, the extent of switching costs depends on how much time and effort a user spent in learning how to use an old product. The more the time and efforts are high, the more the switching costs rise.

For instance, a typical advantage of first movers is that they are able to establish a dominant design which quickly diffuses in the market and imposes to its adopters the effort to learn a unique bundle of knowledge related to its utilization. In this way, the first mover product becomes the technological and "cognitive" standard that its competitors are forced to follow. A similar situation characterized the software industry, where Microsoft with its suite (Microsoft Office) was the first mover in the market segment of office suites. Microsoft launched the program Word 1.0 for Macintosh in 1984 (first year of commercialization of this computer) before packing it with other applications (as Excel or Powerpoint) into the first Office suite 5 years later. All its direct competitors emerged afterward were obliged to design their products taking as benchmark Microsoft Office in order to minimize users' switching costs and any lock-in risk. The lack of switching costs is one of the reasons why nowadays the Internet is replacing quite easily and quickly television in broadcasting: because the utilization of television does not require any particular knowledge or experience and, thus, users have no cognitive or emotional costs in switching to web-broadcasting.

The use of new technology may imply the acquisition for its potential users of new competencies and capabilities. In this case, they have to implement a multistep process of learning through which to acquire codified information and/

---

[13] "Lock-in arises whenever users invest in multiple complementary and durable assets specific to a particular information technology system" (Shapiro and Varian 1999, p. 12).

or tacit knowledge from external environment and social relationships, internalize this new bundle of knowing, and exploit it in order to become able to utilize new technology. The single user absorptive capacity (Cohen and Levinthal 1990) can reduce the impact of switching costs on the potential adopter of an innovation. In organization learning studies, absorptive capacity is commonly defined as "the ability to value, assimilate and apply new knowledge" (Cohen and Levinthal 1990). This notion is relevant also for adoption studies as many times the use of a new technology obliges people to acquire new technological competencies and skills through a complex process of learning. The more people are able to implement this process easily through absorptive capacity, the more they are likely to afford fewer costs in the technological transition.

At micro level, the adoption of an innovation is not solely a problem of economizing but also an issue of feelings and emotions toward the old technology and the new one. The perception of potential adopters about innovation is another critical micro factor influencing adoption. Rogers (1995) identifies five attributes of innovation: relative advantage, complexity, compatibility, trialability, and observability. Compatibility is probably the most important attribute in relation to the issue of technological transition and substitution. The author defines it as "the degree to which an innovation is perceived as consistent with the existing values, past experiences, and needs of potential adopters." Therefore, if an innovation is not compatible with the old product utilized by potential adopters, then it has less chances of being adopted.

Finally, the emotions felt by users toward an old technology can affect their decision to adopt a new substituting product. A recent study on technological change in the mobile phone industry shows that users usually have a strong emotional attachment to their mobile phones (Vincent et al. 2005). It brings about designers to consider carefully this dimension when developing new products and services for this industry. For instance, the feeling of nostalgia felt by some customers for given old products often is the basis for retro-marketing and brand revival strategies (Brown et al. 2003) by which firms propose new (and technologically upgraded) versions of these products to these individuals (as in the case of Volkswagen New Beetle).

If micro factors do not induce individuals to change, they should prefer to use old technology and delay the adoption of new technology.

Summarizing, technological, social, and learning factors are often interconnected to each other and draw a complex picture of adoption, diffusion, and substitution. This picture largely affects the impact and speed of technological change. Old technologies are likely to continue to be used when overall they satisfy users technologically, socially, and cognitively better and more than new technology. In this case, the relevant opportunities for firms may come from arranging some collaborative relationships between old and new technology.

The factors outlined for each category are summarized in Table 1.4.

**Table 1.4**  Main factors commonly associated with user adoption of new technology

| Level | Unit of analysis | Main factors |
|---|---|---|
| Macro | Industry | Network Externalities, Maintenance Services and Distribution Channels, Complementary Technologies and Products, Industry Standard (Dominant Design), Regulations and Policies |
| Meso | Social group | Group cultural and symbolic specificities, Group culture open toward innovation, Group relational specificities, Opinion leaders' orientation about new technology, Social contagion within the group |
| Micro | User | User Absorptive Capacity, Switching Costs afforded by the User, User Perception toward the innovation, User Emotions toward the old technology |

# References

Abernathy WJ, Clark KB (1985) Innovation: mapping the winds of creative destruction. Res Policy 14(1):3–22

Abernathy WJ, Utterback J (1978) Patterns of industrial innovation. Technol Rev 50:41–47

Adner R, Snow D (2010) Old technology responses to new technology threats: demand heterogeneity and graceful technology retreats. Ind Corp Change 19(5):1655–1675

Anderson P, Tushman ML (1990) Technological discontinuities and dominant designs—a cyclical model of technological change. Adm Sci Q 35:604–633

Bagozzi RP, Davis FD, Warshaw PR (1992) Development and test of a theory of technological learning and usage. Hum Relat 45(7):660–686

Barnett WP, Hansen MT (1996) The Red Queen in organizational evolution. Strateg Manag J 17(7):139–157

Bass F (1969) A new product growth model for consumer durables. Manag Sci 15(5):215–227

Brandenburger AM, Nalebuff BJ (1995) The right game. Use game theory to Shape Strategy. Harv Bus Rev July–August:57–71

Brown S, Kozinets RV, Sherry JF Jr (2003) Teaching old brands new tricks: retro branding and the revival of brand meaning. J Market 67(July):19–33

Burt R (1987) Social contagion and innovation: cohesion versus structural equivalence. Am J Sociol 92(6):1287–1335

Burt R, Janicik A (1996) Social contagion and social structure. In: Iacobucci D (ed) Networks in marketing. Sage Publications, Thousand Oaks, pp 32–49

Chari VV, Hopenhayn H (1991) Vintage human capital, growth, and the diffusion of new technology. J Polit Econ 99(6):1142–1165

Christensen C (1997) The innovator's dilemma. When new technologies cause great firms to fail. Harvard Business School Press, Cambridge

Cohen WM, Levinthal DA (1990) Absorptive capacity: a new perspective on learning and innovation. Adm Sci Q 35:128–152

Coleman J, Katz E, Menzel H (1966) Medical innovation: a diffusion study. The Bobbs-Merrill Company, Indianapolis

Cooper AC, Smith CG (1992) How established firms respond to threatening technologies. Acad Manag Exec 6(2):55–70

Davis FD (1989) Perceived usefulness, perceived ease of use, and user acceptance of information technology. MIS Q 13(3):319–339

De Liso N, Filatrella G (2008) On technology competition: a formal analysis of the sailing ship effect. Econ Innov New Technol 17(6):593–610

Dosi G (1982) Technological paradigms and technological trajectories. Res Policy 11(3):147–162

Edgerton David (2006) The shock of the old: technology and global history since 1900. Oxford University Press, New York

Fornell C (1992) A national customer satisfaction barometer: the Swedish experience. J Market 56(January):6–21

Foster RN (1986) Innovation: the attacker's ADVANTAGE. Summit Books, New York

Freddi D (2009) The integration of old and new technological paradigms in low and medium-tech sectors: the case of mechatronics. Res Policy 38:548–558

Gandal N, Kende M, Rob R (2002) The dynamics of technological adoption in hardware/software systems: the case of compact disk players. Rand J Econ 31(1):43–61

Gaskell T (1992) Product revitalisation. Eng Manag J 2(3):145–151

Geels FW (2005) The dynamics of transitions in socio-technical systems: a multi-level analysis of the transition pathway from horse-drawn carriages to automobiles (1860–1930). Technol Anal Strat Manag 17(4):445–476

Geels FW, Schot J (2007) Typology of sociotechnical transition pathways. Res Policy 36:399–417

Geroski PA (2000) Models of technology diffusion. Res Policy 29:603–625

Gilfillan SC (1935) Inventing the ship. Follett Publishing Co., Chicago

Grant RM (1996) Prospering in dynamically-competitive environments: organizational capability as knowledge integration. Organ Sci 7(4):375–387

Grubler A (1997) Time for a change: on the patterns of diffusion of innovation. Technological trajectories and the human environment. National Academy Press, Washington, DC, pp 14–32

Hall B (2004) Innovation and diffusion. In: Fagerberg J, Mowery D, Nelson RR (eds) Handbook of innovation. Oxford University Press, Oxford

Henderson RM, Clark KB (1990) Architectural innovation: the reconfiguration of existing product technologies and the failure of established firms. Adm Sci Q 35:9–30

Howells J (2002) The response of old technology incumbents to technological competition: does the sailing ship effect exist? J Manag Stud 39(7):887–906

Iansiti M (1998) Technology integration. Harvard Business Press, Boston

Klepper S, Simons KL (2000) The making of an oligopoly: firm survival and technological change in the evolution of the U.S. Tire Industry. J Polit Econ 108(4):728–760

Kodama F (1995) Emerging patterns of innovation: sources of Japan's technological edge. Harvard Business Press, Boston

Leonard-Barton D (1992) Core capabilities and core rigidities: a paradox in managing new product development. Strat Manag J 13(Summer special issue):111–125

Macvaugh J, Schiavone F (2010) Limits to the diffusion of innovation. Eur J Innov Manag 13(2):197–221

Majumdar SK, Venkataraman S (1993) New technology adoption in US Telecommunications: the role of competitive pressure and firm-level inducements. Res Policy 22:521–536

Mahajan V, Muller E (1996) Timing, diffusion, and substitution of successive generations of technological innovations: The IBM mainframe case. Technological Forecasting and Social Change, 51(2): 109–132

Murmann JP, Frenken K (2006) Towards a systematic framework for research on dominant designs, technological innovations, and industrial change. Res Policy 35:925–952

Nelson RR, Winter SG (1982) An evolutionary theory of economic change. Harvard University Press, Cambridge

Norton JA, Bass FM (1987) A diffusion theory model adoption and substitution for successive generations of high technology products. Manag Sci 33:1069–1086

Olsen OE, Engen OA (2007) Technological change as a trade-off between social construction and technological paradigms. Technol Soc 29:456–468

Patel P, Pavitt K (1994) The continuing, widespread and neglected importance of improvements in mechanical technologies. Res Policy 23:533–545

Pavitt K (1998) Technologies, products and organization in the innovating firm: what Adam Smith tells us and Joseph Schumpeter doesn't. Ind Corp Change 7(3):433–452

Pinch TJ, Bijker WE (1984) The social construction of facts and artefacts: or how the sociology of science and the sociology of technology might benefit each other. Social Stud Sci 399–441

Ram S (1989) Successful innovation using strategies to reduce consumer resistance. An empirical test. J Prod Innov Manag 6:20–34

Rogers EM (1995) Diffusion of innovations, 4th edn. The Free Press, New York

Rosenberg N (1972) Factors affecting the diffusion of technology. Explor Econ Hist 10(1):3–33

Rosenberg N (1982) Inside the black box: technology and economics. Cambridge University Press, Cambridge

Rosenberg N (1994) Exploring the black box: technology, economics, and history. Cambridge University Press, Cambridge

Sahal D (1985) Technological guideposts and innovation avenues. Res Policy 14(2):61–82

Saviotti PP, Metcalfe JS (1984) A theoretical approach to the construction of technological output indicators. Res Policy 13:141–151

Schiavone F (2011) Strategic reactions to technology competition: a decision-making model. Manag Decis 49(5):801–809

Schiavone F (2013a) Vintage innovation: how to improve the service characteristics and customer effectiveness of products becoming obsolete. IEEE Trans Eng Manag

Schmookler J (1962) Economic sources of inventive activity. J Econ Hist 22:1–20

Schumpeter JA (1912/1934) Theorie der wirtschaftlichen Entwicklung. Leipzig: Duncker & Humblot. English translation published in 1934 as The theory of economic development. Harvard University Press, Cambridge

Schumpeter JA (1942) Capitalism, socialism and democracy. Harper and Brothers, New York

Shapiro C, Varian HR (1999) Information rules: a strategic guide to the network economy. Harvard Business School Press, Boston

Shy O (2001) The economics of network industries. Cambridge University Press, Cambridge

Snow DC (2008) Beware of old technologies' last gasps. Harv Bus Rev (Jan):17–18

Sombart, W. (1930), Die drei Nationalökonomien, Duncker & Humblot, München & Leipzig

Suarez FF, Utterback JM (1995) Dominant designs and the survival of firms. Strateg Manag J 16(6):415–430

Teece D (1986) Profiting from technological innovation: implications for integration, collaboration, licensing and public policy. Res Policy 15:285–305

Teece DJ, Pisano G, Shuen A (1997) Dynamic capabilities and strategic management. Strateg Manag J 18(7):509–533

Tripsas M, Gavetti G (2000) Capabilities and cognition, and inertia: evidence from digital imaging. Strateg Manag J 21(11):1147–1162

Utterback JM (1994) Mastering the dynamics of innovation. Harvard Business School Press, Cambridge

Venkatesh V, Davis FD (2000) A theoretical extension of the technology acceptance model: four longitudinal field studies. Manage Sci 46(2):186–204

Venkatesh V, Morris MG, Davis GB, Davis FD (2003) User acceptance of information technology: toward a unified view. MIS Q 27(3):425–477

Vincent J, Haddon L, Hamill L (2005) The influence of mobile phone users on the design of 3G products and services. J Commun Netw 4(4):69–73

Von Hippel E (1988) The sources of innovation. Oxford University Press, New York

Wejnert J (2002) Integrating models of diffusion of innovations: a conceptual framework. Ann Rev Sociol 28:297–326

# Chapter 2
# Communities of Practice

**Abstract** This chapter reviews the concept of community of practice. The emerging relevance of this form of social aggregation and the dynamics of learning, knowledge sharing, and participation are stressed. In particular, the chapter focuses on the relationships that these social groups have with technology. A technology-based taxonomy of communities of practice is proposed. Furthermore, a desk analysis of the main reactions to technological change is provided. The literature review outlines that technology-based communities often have an ambivalent reaction, between resistance and openness, to technological change if this process implies the substitution of some of their core technological products. The chapter ends by proposing various future research lines on communities of practice, technological products, and the communities' reactions to change.

**Keywords** Communities of practice • Technology in practice • Social construction of technology • Innovation • Ambivalent reaction to change • Knowledge

## 2.1 The Concept of Community of Practice

The concept of community of practice (henceforth: CoP) is fascinating and relatively new. Risen in anthropologic studies (Lave and Wenger 1991), this notion was rapidly applied by scholars in management and organization studies (e.g., Brown and Duguid 1991; Wenger 2000; Wenger et al. 2002). Wenger (1998) summarizes the CoPs' key characteristics by arguing that:

> A community of practice defines itself along three dimensions: (1) What it is about—its *joint enterprise* as understood and continually renegotiated by its members; (2) How it functions—the relationships of *mutual engagement* that bind members together into a social entity; (3) What capability it has produced—the *shared repertoire* of communal resources (routines, sensibilities, artefacts, vocabulary, styles, etc.) that members have developed over time (Wenger 1998).

F. Schiavone, *Communities of Practice and Vintage Innovation*,
SpringerBriefs in Business, DOI: 10.1007/978-3-319-01902-4_2,
© The Author(s) 2014

The Wenger's quotation outlines three core elements of CoPs. First, the set of community practices. The Merriam-Webster dictionary reports practice is "actual performance or application." Practice is something more than simple codified knowledge about a given phenomenon and often requires more complex learning. Apprenticeship is the typical means by which practice is shared and learned within communities. Social interaction between people is critical in both formal and informal contexts to support apprenticeship, learning, and the rise of CoPs. This view is opposite to static interpretations of the learning process. The focus of communities on practice outlines learning processes and dynamics that cannot be detached by a specific context in order to be effective. CoPs, thus, are social structures in which prevails situated learning (Lave and Wenger 1991). Wenger (2000) recognizes that learning is a social process that becomes effective only if CoP balances between core and boundary processes.

Second, participation is another critical process for CoPs. Situated learning, rooted within some specific context, implies that social interaction is a critical component of effective learning. Participation of community should be active and continuous in order to provide effective learning results to people. Mere engagement is not enough. Lave and Wenger (1991) report the case of American meat cutters as an archetypical case of unsuccessful apprenticeship. In this case, the authoritarianism of masters and the workers' perception of this labor as just an unskilled activity and source of little income, affected negatively the learning and performance of community members. The participation in the community activities is critical to people to feel themselves as its members (Brown and Duguid 1991; Handley et al. 2006).

Third, community members share some notions, competencies, symbols, codes, knowledge, routines, and social norms. They are the inputs of practice, apprenticeship, and situated learning and they serve as regulatory mechanisms of social interactions within communities. Every change in this bundle of resources within a CoP depends on the efforts and interest of its members in renegotiating and defining new shared procedures and practices exploiting the renewed resources. For instance, if in one country the scholars of a scientific community consider books and monographs in the national language more important and relevant than papers in international peer-reviewed journals for career upgrades and professional appreciation, then the adoption of a new system of evaluation of scholars (based on papers published in international journals and not anymore on national monographs) can occur only if they decide to change and renegotiate the values, procedures, and practices shared within their scientific community.

Table 2.1 reports the key characteristics of CoPs. The key goal of community members is to share knowledge and information and develop new capabilities of a given practice. This purpose, not oriented to a specific business goal, distinguishes CoPs by all the other typical forms of social aggregation within organizations. Members select themselves autonomously on the basis of their passion, commitment, and expertise about the core practices of the community. Another critical characteristic of CoPs is that they emerge spontaneously from larger informal networks of people with similar working-related interests.

**Table 2.1**  CoPs and other types of social aggregations within companies (adapted from Wenger and Snyder 2000)

|  | Purpose | Members | Links between members | Life length |
|---|---|---|---|---|
| CoP | To develop members' capabilities; to build and exchange knowledge | Members who select themselves | Passion, commitment and identification with the group's expertise | As long there is interest in maintaining the group |
| Formal work group | To deliver a product or a service | Everyone who reports to the group's manager | Job requirements and common goals | Until the next reorganization |
| Project team | To accomplish a specified task | Employees assigned by senior manager | The project's milestones and goals | Until the project has been completed |
| Informal network | To collect and pass on business information | Friends and business acquaintances | Mutual needs | As long as people have reasons to connect |

CoPs could be related to a formal organization or emerge and develop spontaneously. In both the cases members should participate voluntarily. The lack of voluntary participation by members should characterize other types of social aggregation (e.g., project team), different from communities.

Most of the literature focused on the characteristics, dynamics, and implications of formal organization-related communities. A well-known case in the literature of organizational CoP is the Xerox community of technicians, and insurance claims administrators (Orr 1996). Organizational CoP is a heterogeneous form of voluntary social aggregation with specific purposes, procedures of affiliation, links between members, and survival perspectives (Wenger and Snyder 2000). There are various types of organizational CoPs (Wenger et al. 2002):

- unrecognized: these communities are invisible to the organization and sometimes, paradoxically, even to their members;
- bootlegged: only invisible informally to its members;
- legitimized: officially sanctioned as a valuable entity;
- supported: the community receives direct resources (e.g., technological infrastructure, physical spaces) by the organization;
- institutionalized: the organization acknowledges the existence of the community and gives an official status to it.

Typical examples of spontaneous CoPs are informal groups of practitioners sharing a common passion for some craft and skill-based activity (e.g., photographers, snowboarders, and cyclists). The book by Lave and Wenger (1991) is an anthropologic study commonly considered as the first scientific reference on CoPs.

This research describes five informal (not company-related) communities in which apprenticeship and situated learning occur: Yucatec midwives, native tailors, navy quartermasters, meat cutters, and alcoholics anonymous.

Social interaction is the key mechanism for the rise, development, and success of CoPs. However, social interaction itself is not enough. The literature reports other mechanisms to allow planned and spontaneous communities to survive and grow over time. Seven principles make social interaction successful and support the development of effective organizational architectures for this form of voluntary social aggregation (Wenger et al. 2002):

- design and planning of the community evolution over time;
- perspectives from inside and outside the community need to be integrated in order to achieve effective learning and development dynamics;
- invitation of different levels of participation, which should be normally three in every community: a small core group (in which the community coordinators are), an active group, and a large mass of peripheral members;
- development of both private and public community spaces in which community members can meet and exchange ideas and information by many-to-many or one-to-one interactions;
- effective communities must deliver some form of value to its organization or its members. This value very often refers to the benefits and solutions to problems emerging by repetitive meeting between community members;
- combine familiarity and excitement in order to allow members to learn and to get interesting information and new social contacts in an informal and a comfortable environment;
- create a rhythm for the community, for instance, by organizing regular meetings, teleconferences, events, and/or web-activities. All these activities give aliveness to the community and increase people's interest to participate.

These principles facilitate the interaction between the individuals enrolled within a CoP. If it is not enough or successful, all the three key elements suggested by Wenger (1998), namely practice sharing, participation, and development of communal resources, miss and the community will soon disappear.

## 2.1.1 Communities of Practice and Knowledge

A community of people sharing a common practice or passion is something much deeper than a generic social system or segment of product/brand users. A CoP overlaps, to some extent, the notion of market niche. One of the main differences between these concepts is the role of knowledge. This resource binds together people within a community. Conversely, it is less relevant in defining the boundaries of a market niche (which is a concept more related to the strategic analysis of firms). Several studies explore the importance of knowledge within CoPs. Knowledge is critical within these social systems for both apprenticeship and situated learning and (as argued in the next subsection) for development of innovations (Brown and Duguid 1991; Wenger et al. 2002). Three types of knowledge are embedded within every CoP (McLure Wasko and Faraj 2000):

- knowledge as object: justified true belief. Contents of organizational memory including documents and electronic databases. In this sense, knowledge is synonymous of the practice that people share;
- knowledge embedded in individuals: sum of individual knowledge, which is known;
- knowledge embedded in the community: the social practice of knowing. Knowledge existing in the form of routines and shared languages, narratives, and codes.

Amin and Roberts (2008) distinguish four types of "knowing in action" in CoPs: craft/task-based, professional, epistemic/creative, and virtual. These different types of knowing form a typology of knowledge in CoPs (Table 2.2).

In craft-based communities, the relationships between master and apprentices are critical. Apprentices learn the practice and a certain set of related tasks by interacting with their masters within a specific socio-cultural setting. This type of relationship is typical in the manufacturing of artisan/hand-made goods or the development of specialized services. Two specific conditions characterize craft-based

**Table 2.2**  A knowledge-based typology of CoPs (adapted from Amin and Roberts 2008)

|  | Type of knowledge | Social interaction | Innovation | Organizational dynamic |
|---|---|---|---|---|
| Craft/task based | Aesthetic and embodied | Face-to-face communication and co-location, interpersonal trust, long-lived, apprentice-based | Customized, incremental | Hierarchically managed, open to new members |
| Professional | Specialized, declarative, and technologically embodied | Long-lived and slow to change, co-location, institutional trust | Incremental or radical but bound by professional rules | Large hierarchical organizations, restrictions to the entry of new members |
| Epistemic/ creative | Specialized, temporary, and aimed at extending the knowledge base | Spatial proximity, short-lived, based on reputation | High energy, radical innovation | Group/project managed, open to new members with reputation |
| Virtual | Codified and tacit, explorative and exploitative | ICT-mediated interaction, long- and short-lived, weak social ties | Incremental and radical | Carefully managed by moderators. Open but self-regulating |

communities. First, they are primarily concerned to replicate and preserve existing knowledge to produce a given product or service. Second, members of these communities are usually work colleagues sharing a common language, symbols, routines, trust, and interdependencies.

Professional communities are groups of persons engaged in a similar professional activity (e.g., medicine, law, architecture, and consulting). Both tacit and codified knowledge is critical for members of these communities. Formal knowledge is well established and is often a precondition to officially become a member of this type of community. For instance, the learning and respect of ethics codes is mandatory in every professional body. But professional communities make valuable their professional practices and services also by knowing a set of implicit conducts, tacit conventions and competencies. These intangible assets are learned by imitation and repetition of tasks.

Communities of highly expert/creative people are used to develop new exploratory knowledge through collaboration. Creativity is achieved by the integration of people with heterogeneous competencies and expertise. Four factors guide people of CoPs to develop new knowledge and creative solutions to problems (Amin and Roberts 2008): disclosure and peer recognition among creative experts; professional integrity and efforts in collaboration; enthusiastic participation supported by the culture of the interactive setting ("interactive milieu") in which members collaborate; mechanisms of alignment (e.g., codification) between the different actors of the creative process.

Virtual CoPs are the emerging form of practice-based aggregation. Over the last two decades, the Internet supported the aggregation of people communicating and sharing information about their interests, passions, or professional activities. Motivations for knowledge exchange in virtual CoPs are various. McLure Wasko and Faraj (2000) categorize these motivations as tangible returns, intangible returns, community interests, and barriers to participation.

CoPs are informal and spontaneous forms of social aggregation. However, companies should govern organization-related communities in order to exploit the cognitive outputs of their members. An organisational sponsor should constantely work for the achievement of this end. This is a figure of the corporate top management supporting the interaction between the formal organization and the community.

There are several governance mechanisms that organizations can implement to achieve this goal (Probst and Borzillo 2008): stick to strategic objectives, divide objectives into subtopics, form governance committees with sponsors and community leaders, have a sponsor and a CoP leader serving as "control agents," regularly feed the CoP with external expertise, promote the access to other intra- and inter-organizational networks, the CoP leader must have a driver and promoter role, overcome hierarchy-related pressure, provide the sponsor with measurable performance, and illustrate results for CoP members. Conversely, issues that lead to the failure of organization-related CoPs are: lack of a core group within the community, low levels of one-to-one interaction between CoP members, rigidity of competencies, lack of identification with the CoP, and practice intangibility.

## 2.1.2 Communities and Innovation

Innovation is one the most common strategic intents pursued by CoPs.[1] Innovation communities are aimed at developing unexplored ideas and new products or services. Communities of users very often do innovate in order to develop or support some specific products. This behavior is very common in communities of open source software. The development of the browser Firefox is probably the most successful case of a large virtual CoP developing open source software. In this case, the members of the CoP are themselves users of the product that continuously develops and innovates under the co-ordination of Mozilla Foundation. Various motivations explain the reasons for innovation by communities of consumer users (Borgers et al. 2010). First, user innovation entails the exploitation and recombination of knowledge that is not easy to achieve. Conversely, this knowledge and information is sticky (hard to transfer) and locally embedded within the communities. Community members are often interested for professional or personal motives to gain this intangible asset. Second, the expected benefits from selling innovation are another important motive for consumer innovation.

Innovation (in broad terms) is the final output of every organizational CoP. In IBM Global Services, the development of organizational CoPs is planned as an evolutionary process made out of five stages (Gongla and Rizzuto 2001): potential stage, building stage, engaged stage, active stage, and adaptive stage. Each stage characterizes some function performed by the emerging community. The last stage of this evolution (adaptation) supports innovation and generation of new business objects (e.g., new solutions and new services) by communities. These activities support the company to respond to its external environment and allow the development of new capabilities and/or outputs.

The marketing concept of consumer tribes (Cova et al. 2007) is quite similar to the notion of CoP. However, the object of passion of tribes is a specific artifact, good, or brand and not (as for CoPs) the practice performable by it.[2] Some consumer tribes are groups of entrepreneurs[3] as their members develop new product versions or services to share within the community or to sell outside. In this case, tribes could become direct competitors of companies in extracting value from the market.

---

[1] Wenger et al. (2002) report that communities can pursue three other strategic intents: to provide help to members in everyday life (helping communities), to focus on the development, validation, and dissemination of specific practices (best-practices communities), and to organize and distribute knowledge between its members (knowledge stewarding communities).

[2] "Consumer tribes [...] do not consume things without changing them; they cannot consume a good without it becoming them and them becoming it; they cannot consume a service without engaging in a dance with the service provider, where the dance become the service. Participatory culture is everywhere" (Cova et al. 2007 p. 3–4).

[3] The authors develop three other metaphors for consumer tribes: activators, double agents, and predators (Cova et al. 2007).

CoPs support free their members innovators in the development of innovation. A number of motivations explain this behavior (Franke and Shah 2003):

- community-based motives (reciprocal expectations and trust with other community members) for giving support to innovators are more relevant for CoP members than individual-based motives;
- a lack of competition between CoP members;
- the innovation is freely shared and not sold in the community.

Despite these positive evidences, CoPs sometimes might not support and, conversely, make harder the development of organizational innovation. For instance, Swan et al. (2002) in their study on the impact of communities on the development of radical innovation against prostate cancer found that CoP members within can put substantial constraints on radical innovation. These issues relate to work practices and power relations between the subgroups of physicians, technicians involved in the innovation process, and managers. CoP members interact with managers for the development of radical innovation. The nature of these interactions is multilayered and networked and implies the adoption, by companies, of some managerial mechanisms:

1. the construction of a new multidisciplinary community focused on the disease and the proposed innovation. The aim of the emerging community was to find new agreements between heterogeneous groups of medical professionals about treatments for curing prostate cancer;
2. networking and knowledge brokering between the various community subgroups of medical professionals, everyone with different initial ideas on how to cure the disease and the utility of radical innovation to this end. The aims of these actions were to support and facilitate communication and interaction between the subgroups forming the community.

Various scholars (e.g., Wenger et al. 2002; Probst and Borzillo 2008) report the case of DaimlerChrysler, the giant American carmaker, to describe the positive impact and value of CoPs for new product development. Communities of 20–30 engineers and technicians from different car lines collaborate to develop common assembling practices. The name for these communities is "Tech Clubs." These aggregations started informally with regular meetings between former colleagues from functional areas. The key strategic objective of these communities was to develop new efficient assembling techniques to apply uniformly across different technological platforms. In 1996, a critical step to achieve this goal was the creation of the *Engineering Book of Knowledge* (EBoK). This was not a simple textbook on best practices for company engineers. EBoK was a database in which the best practices were summarized but also compliance standards, suppliers requirements, and all information critical to engineers in DaimlerChrysler were documented. Tech Club members exploited EBoK as the opportunity to consolidate their knowledge in designated areas of engineering and promote their role of innovation agents within the organization (Wenger et al. 2002). The rise of Tech Clubs in DaimlerChrysler brings about various practical benefits for the company:

reduction of production costs and time-to-market; increase of productivity (Probst and Borzillo 2008).

Scholars well know the downside of CoPs for innovation. Two basic types of disorders hampering innovation might occur within communities (Wenger et al. 2002). First, a community might not function and miss some of the key requirements of CoPs (e.g., lack of trust between members). This situation undermines the potential and pro-innovation capabilities of community members. Second, disorders making innovation harder in CoPs are related to human frailties. For instance, when communities work too well and their members, thus, take as implicit and unquestioned, many community assumptions. This situation leads them to become "prisoners" to the community routines and traditional mechanisms. In this case, the risk for the community of embracing and preserving an ideal and outdate structure is high. The main weaknesses of this risk are the resistance to change and scarce interest in learning new practices and/or renewing old ones.

These evidences outline that CoPs might respond to technological change in various ways. One variable affecting community reactions is the specific set of relationships between the community and technology.

## 2.2   Communities of Practice and Technology

This section reports various arguments about the connection between technology and CoPs. The first subsection discusses the theoretical approach of the social construction of technology. The second subsection analyzes the meanings and levels of utilization of technology within communities. Drawing on these theoretical arguments, the section ends by providing a typology of communities based on the (central or peripheral) role of technology and artifacts for these social groups.

### 2.2.1   The Social Construction of Technology

Any analysis of the relationship between technology and CoPs should take into account that communities are, at the end, social systems in which a number of different actors interact. The perspective of technological determinism assumes technological evolution drives and orients the evolution and development of society and culture. Over time several scholars from economics of technology, and sociology greatly criticized this approach. Social relationships between social actors, of course, also matter. The evolution, success, and failure of every technology depend on the social interactions, power relationships, and personal and/or collective interests of some actors of these social systems.

The relatively new scientific field of sociology of technology acknowledges and analyzes this last assumption. For some decades a number of scholars in this field does research on the various social dynamics by which technological artifacts are developed, emerge, succeed, or fail in their market diffusion and penetration. The Social Construction of Technology (SCOT) approach[4] by Bjiker and Pinch (1984) is a theoretical framework critical to understand these dynamics. This approach is alternative to the traditional linear approach of technological change and innovation (from basic research to the usage of final goods by customers). Moreover, the SCOT approach can be successfully integrated with the theory of technological paradigms in order to analyze the evolution of socio-technical systems (Olsen and Engen 2007).

The SCOT approach introduces into the analysis of technological evolution the notion of relevant social groups. These groups are usually professional communities sponsoring the technology (e.g., engineers, designers, etc). Social groups refer to institutions, organizations as well as organized or unorganized groups of individuals. The main assumption of the SCOT approach is some relevant social groups support, hamper, and affect the development and adoption of technology and artifacts. A single artifact can bring about different types of specific problems and interpretations for each relevant social group using it. Here lies the meaning of the notion of "interpretative flexibility," the key concept of the SCOT approach.

The technological evolution of this artifact will depend on the mediation of the different relationships between groups. Artifacts are not just technical objects but, using the terminology of Bjiker, they should be termed *socio-technical ensembles*.[5] The development of new technological artifacts is the outcome of a process of alternation between variation and selection. Variation refers to an evolution of the artifact aimed at being the solution of the problem raised by a given social group. Selection is affected by the solutions that each artifact provides to specific consumer problems.

Technological knowledge and its evolution are shaped by specific dynamics, opportunities, and constraints related to the communities of professionals developing the technology. These social groups suggest some technological trajectories, which are mediated by the technological systems embedding it and the functional/organizational decisions of firms (Constant II 1987).

A number of scholars adopted the SCOT approach to analyze, in many other cases, the function and action of social groups for orienting innovation and technological change. Bjiker and Pinch (1984) analyze the case of bicycles. Relevant social groups were critical to speed-up the introduction of automobiles in the rural

---

[4] Social constructionism is a sociological theory of knowledge that considers how social phenomena or objects of consciousness develop in social contexts.

[5] Bijker (1993, p. 124) argues about the notion of socio-technical ensemble that "purely social relations are to be found only in the imaginations of sociologists or amongst baboons, and purely technical relations are to be found only in the wilder reaches of science fiction. The technical is socially constructed and the social is technically constructed. All stable ensembles are bound together as much by the technical as by the social." Therefore, socio-technical ensemble is a notion different and "larger" than the traditional idea of technological artifact.

United States (Kline and Pinch 1996). The SCOT approach was used also in order to analyze the diffusion of electricity in the United States in the 1800s (Geels 2005). Similarly, the community of engineers working in Electricite de France (EDF) was critical for the development and the introduction of the VEL automobile, an innovative electric car, in France (Callon 1987).

This body of literature outlines that technology is not developed just for technical-based reasons. Social forces within social systems (e.g., power relations within the community of designers) influence the extent and direction of such developments and the rate of adoption. These are outcomes of a social battle between different groups of users and developers. The practice exploited by users via technology and artifacts is one key element affecting this social battle.

Referring to CoPs, the SCOT approach outlines two critical implications of the relationship between communities and technology:

1. Communities are not just simple groups of users of some artifacts. CoPs are, by definition, likely to be relevant social groups for the evolution and innovation of their core technological artifacts.
2. Artifacts' evolution is likely to be evaluated mainly in relation to the potential problems that this change could bring about for situated learning and performance of the community's core practices.

## 2.2.2 The Meanings of Technology for Communities of Practice

The SCOT approach provides a general lens for investigation on the relationship between technology and CoPs. This perspective assumes the link is strong, deep, and multilayered. Communities very often are devoted to the use of some socio-technical ensembles. Also, a number of scholars from other scientific fields analyzed the link between communities and technological evolution (e.g., Hoadley 2012). Every community has a common and historical heritage made of goals, beliefs, and stories (Barab and Duffy 2000). All the components of this heritage are somehow related to the utilization, more or less deep, of some technology. People need some set of different technological products to perform a practice (e.g., motorcycles by which drivers enjoy free time on Sunday), provide a repository of information resources (e.g., a wiki-page on Internet in which each driver reports his preferred routes for motorcycling on Sunday), communicate and exchange information and knowledge about the practice itself (e.g., web-chat by people organize in detail their weekly meeting), provide information about the community resources (e.g., PC software by which people share information about relevant community artifacts) (Hoadley 2012).

The variety of technologies useful to CoPs implies different types of knowledge are critical for the CoP survival and practice development by members. Some technological products (e.g., radios or smarthphones) can be core objects for practice implementation in some communities and just means for communication and knowledge exchange in others. On this point, a critical distinction in the organizational literature (Orlikowski 2000) refers to the impact of recurrent interaction between people and technology. Drawn on a structuration-based perspective of technology, the distinction is between two different aspects of technology: technology as artifact and the use of technology-in-practice. The distinction is based on the refusal of the idea that technology and artifacts are stabilized objects across users and organizations. Social action gives to them some specific emergent attributes, properties, and structures making their use particular and not generalizable.

On the one hand, technological artifacts are the goods that we normally use detached by our specified, shared, and repetitive uses and practices. Their use is not particularly meaningful itself for CoP members. However, these artifacts are complementary but not directly involved in the daily practices of the community.

On the other hand, technologies-in-practice are "the set of rules and resources that are reconstituted in people's recurrent engagement with the technologies at hand" (Orlikowski 2000, p. 407). They refer to the use and interaction that people make of technological artifacts. Use of technology implies personal and social experiences of individuals. Technology users and technology structures affect each other. Some properties of the artifact are invisible to users while other properties are wellknown. These structures are relevant only for some goods that people within communities use to perform their core practices. In other words, social action within communities converts some products from simple artifacts into sociotechnical ensembles.

This distinction is consistent with the results of a research program that lasted for more than 20 years at PARC, the Xerox Palo Alto Research Center. The main subject of this broad and long ethnographic study was the design of digital technologies. This activity (such as many others) should be closely linked to the analysis of the sites in which technology is produced and the structures by which it is used. The development of organizational and technological systems, thus, is primarily based on the integrative analysis of social and material specificities related to work and technologies-in-use and on the cultural production of new forms of practice (Suchman et al. 1999).

The SCOT approach also provides a suitable lens to emphasize the key difference between these two aspects of technology. CoPs express problems and find solutions only about goods and artifacts for which they experience some level of technology in practice. Conversely, CoPs are not interested in goods marginal to their practice or learning process (which remain just artifacts) and, thus, cannot be considered as relevant social groups shaping the technological innovation of these objects. This consideration suggests there is a sort of technological hierarchy within each community.

As above argued, there are four basic techniques that people can follow to support CoPs by technology (Hoadley 2012). The first is linking people with others sharing similar practices. The second is to provide access and use of the shared repository of information resources about the practice and its community. The third is the support of communication between community members by technology. The fourth technique is to provide awareness in a community by technology about the information context of various resources.

In sum, these distinctions and techniques outlining technology have various levels of exploitation and meaning in CoPs. However, technology and artifacts have very different meanings across communities. Technology is compulsory for the implementation of practices by every community but its tradition, necessity, and socio-cultural value is variable and changes across communities.

### 2.2.3 A Technology-Based Typology of Communities

The SCOT approach and the distinction between technological artifacts and technology in practice developed by Orlikowski outline a typology for CoPs based on two different types: non-technology-based communities and technology-based CoPs. The key difference between these two types of communities is, as discussed below, the centrality of technology for the practice implementation by their members. These two types of communities should be considered the extreme and opposite points of a continuum. As often occurs for these types of categorizations, many intermediate situations, integrating characteristics of both the ideal-types, could exist in practice. The main evidence for this typology is an important phenomenon as industry technological change does not impact uniformly on CoPs. The extent of its influence depends on the structural characteristics (e.g., practice and social ties) of the community.

Non-technology-based CoPs are groups of people in which technology and physical artifacts have peripheral impact and implications on community practices (e.g., yoga or karate). These groups implement practices that do not require a deep or meaningful use of some tangible and physical technological products. The key elements of their practices are intangible. These types of communities will use some tangible artifacts but they will not give any specific added value or meaning to these objects. Artifacts are just artifacts and no relevant structures or resources are associated to them. For instance, the key practices for a community of yoga practitioners will be based on knowledge about oriental religion and meditation. They use mats to practice but these objects have a marginal impact on the practice development and learning. However, technology can be useful also for such CoPs by providing physical or virtual infrastructures for the community learning dynamics. Technology is critical in these groups only for those activities detached by the direct implementation of the community core practices (e.g., communication and sharing of codified knowledge between people).

This overall marginal relevance of technological artifacts makes scarcely probable any form of user innovation and social construction of technology in these communities. For this reason, they could be defined also as "software-based." However, technological change anyway can impact greatly on the key practices and competencies of such CoPs over time. For instance, a detailed analysis on the US community of pharmacists (Savage 1994) reports the change of the practice in pharmacy prescription due to the improved large-scale manufacture by drug companies over the last two centuries. The key practice of this professional community is the transformation of chemicals into drugs. This practice is based on the pharmacists' knowledge about their patients, chemicals, and medicines (Hibbert et al. 2002). However, a huge technological change in the drug production and distribution occurred along with critical changes in the US laws. In about 40 years (from 1930s to 1970s), the percentage of prescriptions made by physicians requiring the pharmacists' compounding skills rapidly decreased from 75 to 1 % (Savage 1994). One of the reactions to technological change by pharmacists was the revision of their traditional tasks with additional practices beyond the provision of prescribed medications (Harding and Taylor 1997).

The origin of other communities, instead, derives directly from the use of a specific technology or product. These CoPs can be defined as hardware-based and are related to the notion of technology-in-practice. For instance, photographers or artisan woodworkers have to use some specific object (namely, cameras to take pictures and routers for routing out wood) to perform their practices. They are groups of people sharing passions and practices, which are based on the repeated use of some technological product. Their core practice cannot be detached by physical objects which, thus, are central elements of the internal processes of discussion, situated learning, and idea-sharing. These communities, especially if work-related, can be relevant social groups affecting with their opinions, interests, and power the innovation process and evolution of a given technology.

The central role played by technology in this type of community makes the notion of CoP very close to the concept of mass-producing community, a social constellation of actors (organizations and/or individuals) collaborating to produce some technological final goods, sharing a common technological knowledge and usually competing with each other (Preece and Laurila 2003).

The close connection between practice and technology implies that such communities have a strong technical culture and traditions which orient and affect largely the beliefs, knowledge, routines, social norms, and orientation to innovation of their members. Constant II, in his work on turbojets (1987), finds that technological knowledge refers to traditions of practices possessed by communities of technological practitioners. The author finds that design and production of turbojets are processes based on traditions shared within the communities of designers and engineers. These professionals have to extend and articulate their technological traditions in order to innovate incrementally their practice. Constrains to the incremental development of traditional community practices might be the risk of functional failure and the incapacity to function under new conditions.

Of course, not every artifact will be based on some technology-in-practice for the members of a hardware-based CoP. People in these communities will use also artifacts not directly related to the implementation of their core practices or apprenticeship. For instance, artifacts having some value in terms of technologies-in-practice for bicyclers are bicycles and their wheels, whereas simple technological artifacts are ride gloves or helmets. Bicyclers use all these products but it is evident that they have very different impacts on their practices and performances.

In technology-based CoPs situated learning and apprenticeship involve mainly the use and practice of technological artifacts. Physical objects, therefore, affect

**Table 2.3**  Communities of practices and technology

|  | Hardware-based | Software-based |
|---|---|---|
| Example | Photographers, snowboarders | Yoga practitioners, Karate practitioners |
| Meaning of technology | Technology in practice | Just technological objects |
| Situated learning | Learning strongly linked to technological evolution | Learning decoupled by technological evolution |
| Social interactions | Centered around some core technologies | Supported by (but not centered around) technology |
| Social construction of technology | Strong impact | Weak impact |
| Development of technological innovation | Probable | Not probable |

deeply all the typical key processes of situated learning in communities as reported by Handley et al. (2006):

- participation: objects are some of the key reasons for which people participate in community;
- identity: the expertise and use of objects have a symbolic value for community members. Artifacts contribute to build or provide a specific and recognizable identity to the community;
- practice: this activity cannot be detached by the recurrent use of technological artifacts. This property makes such communities focused also on the practice-oriented use of artifacts and not just to the artifacts themselves.

The centrality of technology for practice and learning makes very probable the development of some technological or service innovation in such communities. Table 2.3 summarizes the basic differences between the two types of CoPs described in this subsection.

## 2.3  Reactions of Communities of Practice to Technological Change[6]

Technological change can bring about various reactions by technology-based or hardware-based CoPs. Change is an endogenous and inevitable process for the community technology in practice. These changes depend on human actions and interactions.

Practices related to these objects can be stabilized even for a long time but they will evolve over time as a consequence of various external unpredictable conditions (e.g., evolution of technology, preferences of users and designers) (Orlikoski 2000). Technology evolution is one of the main factors (the others are changing

---

[6] This section is based on and extends some elements of the analysis developed by the author in Schiavone (2012).

market conditions and organizational structures) able to lead communities to radical transformations or even to death (Wenger et al. 2002). Conversely, technological change has a limited impact on the core practices of non-technology-based communities.

In general, technology evolution has a critical impact on the internal dynamics of CoPs. Wenger et al. (2002) argue on this point that:

> Changes in the core science or technology of a community constantly reshape it, often bringing in professionals from neighboring disciplines or introducing technological advances that change their way of working. Because communities are built on existing networks and evolve beyond any particular design, the purpose of a design is not to impose a structure but to help the community develop (p. 53).

In theory, community reactions to technological change can be of resistance or openness (acceptance and adoption) of new technology. In practice, most technology-based communities have ambivalent (hybrid) responses in order to survive and face effectively technological change. The remainder of this section reviews these alternatives.

## 2.3.1  Resistance

Piderit (2000) defines resistance to a change as the set of responses to change that are negative along various critical dimensions affecting the individual action. The author argues that these dimensions relate to the emotions, cognition, and intentions of organizational actors.

Scholars have provided various explanations and described various sources of resistance to new technologies in communities. As assumed in the SCOT approach, technological change is deeply affected by a process of social construction between the main social groups of a socio-economic system (Olsen and Engen 2007). In this view, every new technology is the final result of a process of negotiation between these social groups representing different interests and problem definitions. If a new technology does not fit the interests or values of a given social group, their members will be likely to avoid its adoption. Buhl (1974) analyzed in-depth the resistance to technological change in the American navy between 1865 and 1869 (transition from sailing ships to steamboats). The author concludes his analysis by arguing that "technology is in part a pawn in the grand chess game of social conflict [...] Sponsorship of and resistant to technology both function to secure interests" (Buhl 1974, p. 727). In this view, resistance to technological change occurs because community members have a social interest in preserving the old technology: i.e., the adoption of a new technology could weaken the CoP's identity, tradition, social ties, and culture.

Symbolic and social meanings associated with new technology are critical drivers that support (or hamper) technological change inside communities. The case of the initial contrast by American farmers against cars at the begin of the 1900s shows that if new technology is perceived as risky or unsafe, then members can contrast drastically its adoption and diffusion within a community or social group.

The initial anti-auto crusade of this community against the early models of this artifact (nicknamed "the devil wagon"), basically due to its dramatic effects on rural livestock, changed thanks to the masculine social meaning that cars received. Men farmers started to use cars to perform their "men's jobs" and express their technical competences. Therefore, the emerging masculine vocation of auto-vehicles supported their diffusion within this community (Kline and Pinch 1996).

The "blindness" explanation about resistance to change in old technology-based CoPs is based on the assumption that the practice inhibits the perception and recognition of community members about the improvements offered by new technology. The cognitive universe of the community blinds its members and does not allow them to recognize the utility of alternative technologies to implement their conventional practice (Constant 1984).

Another typical source of resistance to technological change in the field of professional community is the perceived de-professionalization or routinization that innovation could bring about.

Members of a technology-based CoP are likely to be less attracted by innovation as network externalities related to the old product continue to be very intensive within their community although technological substitution is occurring in the rest of the market. The more a community is old and large and its members are interconnected to each other, the more network externalities act as a "shield" making the substitutive innovation less worthy of interest and balancing the possible shortcomings, generated by the new industry configuration, of complementary assets for the old product. For instance, CoPs members often are able to build components and/or arrange complementary services for their old products by themselves, without the need of external distribution or maintenance channels.

A further element generating resistance should be the type of technological artifact, which has to be replaced by technological change. For instance, a dominant design is a product likely to have some type of technology in practice for communities. These widely adopted products can be objects of cult and passion with specific structures (technology-in-practice) for some groups of individuals performing some specific activity by them.

All these considerations suggest CoPs are likely to experience a general resistance toward technological change and innovations substituting their core products due to a number of different types of motivations. Drawing on a recent theoretical model about nonadoption of new technology (Schiavone and McVaugh 2012) and tailoring it to the specific issues of CoPs, three different types of factors are likely to impact positively and increase the resistance to technological change in the members of this type of social group (Fig. 2.1):

- macro factors, directly related to the technological and industrial conditions of the core products of the community;
- meso factors, directly related to the social dynamics and interactions between the community members;
- micro factors, directly related to the personal characteristics and orientations of the single members of the community.

**Fig. 2.1** CoPs and reactions
to technological change

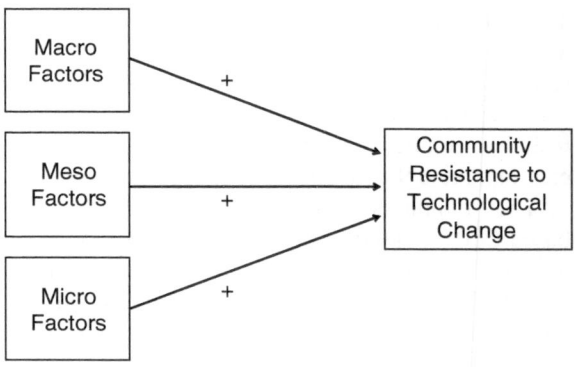

Old technologies and products continue to be used in some CoPs, albeit there is technological change in the rest of the market as they are essential to keep the value of the symbolic and social specificities of their CoPs. Technological change and old technology substitution imply the change in the constitutive knowledge, social norms, and practices of a hardware-based CoP and the learning of new knowledge and practices. Thus, innovations are often likely to be perceived as "risky" for the CoP survival as they may radically change community constitutive routines (the habit) and/or may be socially disapproved by the other members. This brings about technological substitution in CoPs which is likely to be much slower and harder than in other market niches.

The same users might be open to technological change and substitution in order to satisfy the same need in their normal life (e.g., if I belong to a CoP of drivers of vintage cars, I will drive my "Ford Model T" to participate to amateur meetings but, probably, every day I will go to my office driving my modern car). Moreover, the impact of the various levels of factors on CoP resistance to technological change is not homogeneous within a specific CoP for instance, in some CoPs micro factors can be more relevant than meso and macro factors in hampering substitution, whereas in other CoPs meso factors might the key motivation for the resistance. The extent of the impact of each type of factors changes in relation to the nature and type of practices around which the community is built.

In sum, these CoPs present an inner "dark side" of innovation and technological change. CoPs, and more in general niches, are not the sole loci of innovation. CoP members reject and are closed toward radical innovations replacing the dominant artifacts of their community. As reported in Chap. 5, CoP members might react to threats substitutive innovations by creating another innovation by themselves. Indeed, if technological change endangers the survival of a product-based CoP, their skilled members could develop innovations in order to keep alive and reinforce the practices, meanings, and routines of their community.

## 2.3.2  *Openness*

Although technology-based CoPs are motivated to resist technological change, these communities might even perceive the utility of new technology (albeit competing with old one) and be very open to it. In general, small social groups are considered positive environments for innovations. Geels and Schot (2007, p. 401) stressed the criticality of niches for technological change by introducing the concept of technological niches, "small networks of actors [that] support [radical] novelties on the basis of expectations and visions" and act as "incubation room" by protecting novelties against mainstream market selection.

A basic assumption of the literature is that technology and communities influence and co-evolve with each other. In their study of the flight simulation industry, Rosenkopf and Tushman (1998) found that technology and communities of technological users co-evolve. When new technology is in its "era of ferment," dynamics of social construction occur inside the community. When a dominant design emerges and a phase of incremental change begins, technological determinism prevails in the community.

The intrinsic orientation of CoPs to learning makes these social groups suitable aggregations for the adoption of new technology. A number of cases in the history show communities had positive reactions to technological change despite this process implying critical modifications to their routines and competencies. For example, after World War I, the widespread diffusion of aluminum in the aeronautical engineers' community depended upon their association of this metal with the idea of progress in airplanes manufacturing (Schatzberg 2003). Conversely, wood (the incumbent technology) started being perceived as inadequate material for such industrial production. Another interesting case is what happened between tennis players over the 1960s and 1970s, years during which wood rackets, the first dominant design of this community of users, were substituted by metal and graphite-reinforced rackets. Tennis players accepted these technological changes without any resistance as such innovations improved greatly their performances and skills (e.g., by reducing the vibration when the ball is hit). New rackets brought about within the community new competitive techniques and styles of play, as the two-handed backhand. The rising professionalism of this sport and economic prizes for winners of tennis tournaments over that period were critical conditions to support community openness to new technology (Galenson 1993).

Some members shall promote new technology and products within the community to achieve diffusion of innovation and openness to technological change. These members perform the same function as "change agents": i.e., actors aimed at managing transition and leveraging resistance to change in organizations. These individuals are "those who are responsible for identifying the need for change, creating a vision and specifying a desired outcome, and then making it happen" (Ford et al. 2008, p. 362). Technology stewards emerge within communities to exploit and adjust the new technology to the needs of their social group. This role configures an array of technological support that enables the community to function (Wenger et al. 2005). Both single members and community subgroups can perform this function. Three main moments of "inventiveness" characterize the technology steward:

1. inventiveness of the technology market: new technology emerges into the market and some community members perceive it as source of new opportunities for their community;
2. inventiveness of serving the community perspective: new technology is tailored to the community activities and practices;
3. inventiveness of use: new technology understands and meets the latent needs of the community.

People open to technological change often understand its value for rejuvenating the community and its practices. Every CoP crosses normal life cycles. Technological change might provide not just risks but also opportunities for the entry within the community of new members, knowledge, and competencies.

### 2.3.3  Reasons for Ambivalent Reactions

CoPs have specific reasons to both hamper and support technological change. Table 2.4 summarizes these conditions. The final outcomes of this hybrid picture are very often ambivalent responses to technological change by community members. Ambivalence is the intermediate condition output of both openness and resistance to technological change within the community. Ambivalence is likely to come out after the alternation of different reactions changing over time. In the short term, communities can contrast innovation and technological change as soon they are launched. In the long term, community members slowly tend to adopt new technology and integrate it in their traditional practices. This change inevitably brings about a modification of the practices, social interactions, and learning themselves. The case of automobile in US farmers clearly reports this alternation of reactions over time.

In some cases, communities of technological users try to optimize the adoption of innovation by "re-inventing" new technology. Re-invention refers to "the degree to which an innovation is changed or modified by a user in the process of its adoption and implementation" (Rogers 1995, p. 174). For instance, anesthesiologists specializing in anesthesia for cardiac surgery tend to tailor the new technology to their needs, as when they use new and sophisticated computer systems that might create more burdens and complexities (Cook and Woods 1996). Reinvention of new technology is a typical ambivalent reaction as it merges the adoption of new artifacts with the preservation of traditional practices.

CoPs tend to resist technological change if new technology substitutes their old technology-based "core" artifacts. However, typical community phenomena as reinvention or technology stewardship suggest CoP members can exploit new

**Table 2.4**   Pros and cons for technological change within communities

| Reasons for resistance | Reasons for openness |
| --- | --- |
| Strong technical culture about the declining technology | Interest and orientation for experimentation and innovation |
| Preservation of traditional practices, knowledge, symbolic meanings and competencies | Need for survival and integration of internal and external dynamics |
| High switching costs (e.g., learning efforts, emotional constrains) | Presence of change agents supporting new technology within the community |
| | Positive social construction of new technology within the community |

technology in order to adapt to industry technological change and, somewhat, to preserve the traditional implementation of their old technology-based practices.

Ambivalence is a proxy of the fact that within the same community some different subgroups with different opinions on how to react to technological change could emerge. The view of Wenger and his co-workers about a "joint enterprise" of all community members is too optimistic. Various tensions between proponents of alternative reactions within the community makes much more complex its management.

As reaction to these phenomena firms may develop innovations not competing and substituting old declining products but extending their cycle-life by adapting them to the ongoing technological change and connecting them to new technology. These innovations should be marketed to those niches or segments of consumers resisting, for some reason, the substituting innovation.

## 2.4   Research Agenda About Communities of Practice and Technological Change

The literature outlines a number of quantitative and qualitative questions, still unexplored by scholars, about the impact of technological change on technology-based CoPs. These questions may be analyzed on the basis of two different dimensions: (1) the two basic reactions to technological change by CoPs; (2) the distinction between technological artifacts and technology-in-practice. The basic assumption to elaborate further inquiries by these dimensions refers to the CoPs' heterogeneity in their reactions to technological change and set of their technological artifact. Table 2.5 reports the research questions discussed in the remainder of this section.

Referring to resistance to technological change, various future inquiries should refer to minor artifacts used by community members. For instance, a critical question in this field should be the analysis of what types of technological artifacts effectively hamper technological change in hardware-based CoPs. Such as people, no products are islands. Every technological product, big or small, complex or

**Table 2.5**  Matrix for the analysis of technological change within Communities of Practice

|                         | Resistance to change                                                                                   | Openness to change                                                                                                                               |
| ----------------------- | ------------------------------------------------------------------------------------------------------ | ------------------------------------------------------------------------------------------------------------------------------------------------ |
| Technology-in-practice  | What is the real impact of substitution of core products on resistance to technological change?         | How do change agents tailor new technology to support and improve traditional practices, learning processes, and social interactions?             |
| Technological artifact  | What types of minor artifacts hamper technological change?                                              | How does new technology support the survival of a community by providing new complementary products?                                             |

simple is designed to work within some specific technological system. In the case of technological artifacts for technology-based CoPs, the position of these products will be marginal within the technological system relevant for such communities. It should be relevant to understand if and how the linkages between these artifacts and community core products influence somehow people reactions to technological change. What types of minor community artifacts hamper technological change. Moreover, how does the substitution of minor artifacts affect the general reation to technological change by technology-based communities?

Technological change basically brings about critical implications and resistance for the substitution of technology-in-practice used by community members. This point should be elaborated further by scholars especially in relation to the issue of situated learning (by definition these technologies cannot be detached by learning processes within CoPs) and participation to community. Therefore, another set of future inquiries should be about how and to what extent the process of situated learning really affect the contrast to technological change within communities. The relevant questions in this domain could be: does the level of participation to community learning processes matter to this resistance or is it generalized across different levels of engagement? How do social mechanisms and apprenticeship protect technology-in-practice from technological substitution? What are the levels of knowledge most relevant (hold by the organization, individuals, or the community itself) avoiding the replacement of technology-in-practice and the change in social dynamics by which practice is shared?

The answers to these questions would support companies, communities leaders, and sponsors in planning in detail and ex-ante how to overcome resistance to new technology which, instead, could support the rejuvenation and survival of the community itself over time.

Referring to openness to technological change, the extant literature on the key dynamics and characteristics of CoPs outlines a number of interesting new potential research lines. They should be focused on the exploitation of new technology, now accepted by users, for the development or upgrade of minor technological artifacts. The rise of new virtual forms of communities shows that technological change is not an issue for people if it refers to the minor artifacts of the infrastructures of their community. New technology can provide community new complementary tools and instruments for the development and improvement of performance of their core practices. These new artifacts can impact somehow the overall social networks and actions of people, for instance by profiling new ways of performing the community practices. Drawing on these evidences, a critical set of questions in this domain should be: how do new technology-based complementary artifacts improve the performance and make easier learning and practice execution within communities? How much does the adoption of new technologies reshape social interactions or create new subgroups of users within the community? How much technological knowledge about new technology must be spread within the community in order to achieve these goals?

Research on openness to technological change within hardware CoPs should provide relevant findings also if applied to technology-in-practice. When technological evolution refers to the core products of the community, higher barriers toward openness to change should occur. This orientation, thus, should be supported by some community members serving as change agents. These actors should plan and carefully work on how tailoring as best as possible new technology on traditional community practices, social mechanisms, and learning processes. Drawing on this evidence a set of interesting and relevant questions should be: how change agents should hybridize new technology with old technology-in-practice in order to orient effectively their community toward ambivalence and improve practice performances? What elements of the traditional old technology-based practice should be unchanged to involve successfully the other members of the community in the process of adoption of new technology? How much new technological knowledge must be spread and learned by people within the community in order to change consistently the overall core practices and social interactions over time? What are the probabilities of attracting new members to the community by revolutionizing traditional practices with new technology?

A number of community-related variables could affect the research designs and outcomes of these future studies. For instance, these questions can bring about different replies if the analysis is based on work-related or spontaneous CoPs. The degree of institutionalization of a CoP might influence both the general orientation of its members to technological change and the role played by marginal and core technologies. Moreover, another variable impacting on the design and outcomes of these possible future studies might be the stage of the life cycle of the community. Orientation to change and openness to new technology should be greater in young, emerging CoPs.

# References

Amin A, Roberts J (2008) Knowing in action: beyond communities of practice. Res Policy 37(2):353–369
Barab SA, Duffy T (2000) From practice fields to communities of practice. Theor found learn environ 1(1):25–55
Bijker WE (1993) Do not despair: there is life after constructivism. Sci Technol Human Values 18(1):113–138

Bogers M, Afuah A, Bastian B (2010) Users as innovators: a review, critique, and future research directions. J Manage 36(4):857–875

Brown JS, Duguid P (1991) Organizational learning and communities-of-practice: toward a unified view of working, learning, and innovation. Organ Sci 2(1):40–57

Buhl LC (1974) Mariners and machines: resistance to technological change in the American Navy, 1865–1869. J Am Hist 61(3):703–727

Callon M (1987) Society in the making: the study of technology as a tool for sociological analysis. In: Bijker WE, Hughes TP, Pinch TK (eds) The social construction of technological systems. MIT Press, Cambridge, pp 83–103

Constant EW (1984) Communities and hierarchies: structure in the practice of science and technology. Laudan R (ed) The nature of technological knowledge: are models of scientific change relevant?. Riedel, Dordrecht

Constant EW (1987) The social locus of technological practice: Community, system, or organization.? In: Bijker WE, Hughes TP, Pinch TJ (eds) The social construction of technological systems: new directions in the sociology and history of technology. MIT Press, Cambridge, pp 223–242

Cook RI, Woods DD (1996) Adapting to new technology in the operating room. Hum Factors 38(4):593–613

Cova B, Kozinets RV, Shankar A (2007) Consumer tribes. Butterworth-Heinemann, Oxford

Franke N, Shah S (2003) How communities support innovative activities: an exploration of assistance and sharing among end-users. Res Policy 32(1):157–178

Ford JD, Ford LW, D'Amelio A (2008) Resistance to change: the rest of the story. Acad Manag Rev 33(2):362–377

Galenson DW (1993) The impact of economic and technological change on the careers of American men tennis players, 1960–1991. J Sport Hist 20(2):127–150

Geels FW (2005) The dynamics of transitions in socio-technical systems: a multi-level analysis of the transition pathway from horse-drawn carriages to automobiles (1860–1930). Technol Anal Strateg Manag 17(4):445–476

Geels FW, Schot J (2007) Typology of sociotechnical transition pathways. Res Policy 36:399–417

Gongla P, Rizzuto CR (2001) Evolving communities of practice: IBM Global Services experience. IBM Syst j 40(4):842–862

Handley K, Sturdy A, Fincham R, Clark T (2006) Within and beyond communities of practice: making sense of learning through participation, identity and practice*. J Manage Stud 43(3):641–653

Harding G, Taylor K (1997) Responding to change: the case of community pharmacy in Great Britain. Sociol Health Illn 19(5):547–560

Hibbert D, Bissell P, Ward PR (2002) Consumerism and professional work in the community pharmacy. Sociol Health Illn 24(1):46–65

Hoadley C (2012) What is a community of practice and how can we support it? In: Jonassen DH, Land SM (eds) Theoretical foundations of learning environments, 2nd edn. Routledge, New York, pp 287–300

Kline R, Pinch T (1996) Users as agents of technological change: the social construction of the automobile in the rural United States. Technol Cult 37(4):763–795

Lave J, Wenger E (1991). Situated learning: legitimate peripheral participation. Cambridge University Press, Cambridge

McLure Wasko M, Faraj S (2000) It is what one does: why people participate and help others in electronic communities of practice. J Strateg Inf Syst 9(2):155–173

Olsen OE, Engen OA (2007) Technological change as a trade-off between social construction and technological paradigms. Technol Soc 29:456–468

Orlikowski WJ (2000) Using technology and constituting structures: a practice lens for studying technology in organizations. Organ Sci 11(4):404–428

Orr JE (1996) Talking about machines: An ethnography of a modern job. IRL Press an imprint of Cornell University Press, Ithaca

Piderit SK (2000) Rethinking resistance and recognizing ambivalence: a multi-dimensional view of attitudes towards organizational change. Acad Manag Rev 25:783–794

Pinch TJ, Bijker WE (1984) The social construction of facts and artefacts: or how the sociology of science and the sociology of technology might benefit each other. Soc Stud Sci 14:399–441

Preece D, Laurila D (2003) Technological change and organizational action. Routledge, London

Probst G, Borzillo S (2008) Why communities of practice succeed and why they fail. Eur Manag J 26(5):335–347

Rogers EM (1995) Diffusion of innovations. The Free Press, New York

Rosenkopf L, Tushman ML (1998) The co-evolution of community networks and technology: lessons from the flight simulation industry. Ind Corp Change 7:311–346

Savage DA (1994) The professions in theory and history: the case of pharmacy. Bus Econ Hist 23(2):130–160

Schatzberg E (2003) Symbolic culture and technological change: the cultural history of aluminum as an industrial material. Enterp Soc 4(2):226–271

Schiavone F (2012) Resistance to industry technological change in communities of practice: the "ambivalent" case of Radio Amateurs. J Organ Change Manage 25(6):784–797

Schiavone F, MacVaugh JA (2012) Non adoption of new technology. In: Ran B (ed) Contemporary perspectives on technological innovation, management and policy, vol 2. Information Age Publishing, Charlotte

Swan J, Scarbrough H, Robertson M (2002) The construction of communities of practice in the management of innovation. Manage Learn 33:477–496

Suchman L, Blomberg J, Orr JE, Trigg R (1999) Reconstructing technologies as social practice. Am Behav Sci 43(3):392–408

Wenger E (1998) Communities of practice: learning as a social system. Syst Thinker 9(5):2–3

Wenger EC, Snyder WM (2000) Communities of practice: the organizational frontier. Harvard Bus Rev 78(1):139–146

Wenger E, McDermott R, Snyder W (2002) Cultivating communities of practice: a guide to managing knowledge. Harvard Business School Press, Cambridge

Wenger E, White N, Smith JD, Rowe K (2005) Technology for communities. In: CEFRIO (ed) Guide to the implementation and leadership of intentional communities of practice. Work, learning and networked, pp 71–94

# Chapter 3
# Vintage Innovation

**Abstract** This chapter offers the notion of vintage innovation, an innovative approach to improve the customer effectiveness of old products without changing their technical characteristics. The chapter reviews the Saviotti and Metcalfe theoretical framework in order to analyze the key components of technological products (technical characteristics and service characteristics). Backwards compatibility provides interesting opportunities to improve customer effectiveness to date scarcely considered by firms. This form of technological compatibility leads to the phenomenon of vintage innovation. This shows that companies have to focus, paradoxically, their R&D efforts on new technology in order to improve customer effectiveness of declining products. In particular, vintage innovation generates value for companies when users form a community of practice. The chapter ends with the main managerial implications of vintage innovation.

**Keywords** Vintage innovation • Technical characteristics • Service characteristics • Customer effectiveness • Backwards compatibility • Converters

## 3.1 Introduction

The main conclusions of Chaps. 1 and 2 can be merged and summarized as follows: technological change can bring about unexpected strategic reactions by old technology-based communities of practice if this process impacts on their core technological artifacts. In this case, the knowledge, routines, and practices of communities can hamper technological substitution and, as a consequence, lead to unpredic "ambivalent" outcomes. New technology to be effective for and adopted by these groups of users has to contribute to the survival of both the incumbent technology itself and the social practices and structures related to old artifacts.

---

This chapter is based on and extends some elements of the analyses and researches developed by the author in Schiavone (2013a) and Schiavone (2013b).

F. Schiavone, *Communities of Practice and Vintage Innovation*,
SpringerBriefs in Business, DOI: 10.1007/978-3-319-01902-4_3,
© The Author(s) 2014

   In other words, these conclusions suggest the future of such groups of old technology aficionados will be necessarily based on continuous future replications of their past. This view is not new at all in social sciences but finds its roots in the works of various philosophers arguing that inevitably the past comes back into the future of humans. The Italian philosopher and historian Gian Basttista Vico, who lived between the seventeenth and eighteenth centuries, developed the theory of "historical cycles"[1] within his masterpiece *The New Science* (Vico 1725), in which it is proposed that over time some critical events tend to be repeated by human beings. Alexandre Kojève, a Russian-French philosopher who lived in the first half of the 1900s (1902–1968), argued that the society already saw the "end of the history" (Kojève 1980) and in the future human beings will just see replications of already occurred events. Following Kojève's thesis, Fukuyama (1992) develops further this idea.

   The history of technology is rich with cases of recurrences of the past into the future (Edgerton 2006). For instance, the case of AGA cooker shows how some old products acquire a "patina of retro-chic" (Edgerton 2006, p. 57) and users continue to use them even after many years they were discontinued by main companies. The technology management literature outlines very often that modern replications of past technologies are possible just because old technology is able to evolve and provide better performance to face the evolution of external conditions over time (e.g., Adner and Snow 2010). Indeed, such recurrences are technically based on standard techniques, as revitalization, by which companies improve the performance and effectiveness of old technological products. These approaches make old objects persistent, undeletable, endless means for the social construction and practice by some groups of users. In other words, these approaches let the past matter and come back also in the field of technology, and not just for human events as theorized by philosophers as Vico, Kojève, and Fukuyama.

   The notion of vintage innovation (Schiavone 2013a), as described in the remainder of this chapter, shows much stronger the persistence and continuous replication of the past also in the field of technology. A past, so strong and deep for some customers, that cannot be technically revitalized or artificially upgraded. A past that must be today as it was yesterday. This theoretical argument can lead to relevant developments in the field of innovation management. In his crucial work on the principles of entrepreneurship and innovation, Peter Drucker (1985) highlighted seven main sources for innovative opportunities that new firms and/or potential entrepreneurs should monitor: the unexpected, incongruities, process needs, industry market and structures, demographic changes, changes in public perception, new technology, and scientific findings. Vintage innovation assumes that the past can be, extending Drucker's taxonomy, the eightieth source of innovation for companies. The key specificity of this type of innovation is the improvement of technical performance and customer effectiveness of old technological products after technological change without any change in their original technical characteristics.

---

[1] In Italian, the native idiom of Vico, the theory is named "corsi e ricorsi storici."

Chapter 2 reports various factors that may affect negatively the predisposition of some technology-based communities of practice to adopt a new technology or even revitalized versions of old products. Practice execution and the preservation of traditional knowledge and structures are some of the main community needs outlining the possibility of developing a new approach, not based on revitalization, for the improvement of old technological products.

The remainder of this chapter is organized as follows: the next section describes in detail are the types of characteristics (service and technical) of technological products. Section 3.3 proposes the notion of vintage innovation and outlines its main characteristics and differences with the existing techniques for the improvement of performance of old technological products. Section 3.4 describes the main organizational domains related to this approach: technology management, innovation strategy, and marketing.

## 3.2 A Third Way to Improve Old Technological Products?

This section outlines the theoretical foundations of a third way to improve the performance and customer effectiveness of old technology-based products (henceforth: OTBPs) becoming obsolete after technological change. These goals are commonly achieved by two traditional approaches: old product revitalization and retrofitting (variant of the revitalization).[2] Both these approaches inevitably lead to some incremental or architectural innovation of the OTBP. Revitalization leads to important incremental product innovations of the OTBP and implies changes even in its market positioning. For instance, when quartz watches emerged in 1969 and replaced mechanical watches in the mass market, some firms recognized there was a segment of consumers still preferring old-technology watches. These watchmakers just revitalized their old mechanical watches by changing their design and retrenched in this new market niche, even more profitable than their traditional mass market. Retrofitting refers to "the introduction of new technologies into existing product architectures" (Prencipe 2003, p. 126). The basic difference between these two approaches is that pure revitalization companies change some technical characteristic (inner mechanisms and/or design) of the old product to improve its performance without merging it with new technology. If this merge occurs, then the old product has been retrofitted with new technology. Typical examples of retrofitting are hybrid vehicles (merging internal combustion and electricity-based motors) and last generation home telephones (merging analog telephony and digital telephony).

Both these approaches achieve improvements in technical efficiency by some changes in the internal and/or external technical structure of the old product. As a consequence, technical efficiency leads to more technical performance. However, improvements in performance of technological products can also be unrelated to technical

---

[2] Chapter 1 reports a wide description about old product revitalization.

efficiency and lead anyway to gains in effectiveness for customers. The effectiveness of a product refers to "the degree to which the product meets the targeted needs of the customer (i.e., benefits and costs of the product)" (Madhavan and Grover 1998).

These considerations are critical for the arguments in this chapter. Indeed, technical changes might not be appreciated by members in technology-based communities of practice and bring about resistance to innovation. For instance, technical changes in the OTBP might hinder some primary goals of communities facing technological change: the preservation of their traditional practice, identity, structures, resources, and knowledge associated to their core OTBPs. Would the fans of film technology be happy in using a hybrid easy-to-use camera? Maybe not, if these users have relevant expertise, knowledge, passion for analog cameras and are "emotionally involved" with film technology. However, community members might be very interested (or obliged in some cases) in improving the customer effectiveness (benefits) of the traditional versions of their products after technological change. For instance, aficionados of film technology might be curious of experiencing the benefits of digital imaging.

These arguments outline the foundations for an unconventional way to improve the performance and customer effectiveness of OTBPs. The next subsection reports a key theoretical model in the technology management literature to better understand the relationships between the technical features, performance, and evolution of products and a further third way consistent with the ambivalent needs of technlogy-based communities of practice: the Saviotti and Metcalfe (1984) model. The second subsection describes the most suitable technical means for the implementation of this third way: adapters and converters.

### 3.2.1 The Saviotti and Metcalfe Model

Saviotti and Metcalfe (1984) provide an interesting framework to better understand (a) the set of connections between OTBPs' technical efficiency, service characteristics,[3] and effectiveness for customers and (b) the weakness of a restricted (just efficiency-based) view of improvements of performance of OTBPs. Saviotti and Metcalfe focus their work on the evolution of technological products. These authors argue that every new technological product is a sort of "compromise" between some old technologies and some innovative scientific and/or technological advancement. Every product can be described as the combination of various technical characteristics (its inner workings) and various services characteristics (its performance).[4] About it, Castaldi et al. (2009) wrote that:

---

[3] From this point the terms "performance" and "service characteristics" are used interchangeably in the chapter.

[4] Saviotti and Metcalfe (1984) consider also a third set of product characteristics (irrelevant for the present analysis) in their article: process characteristics, referring to the process by which every product is produced.

Technical characteristics represent the internal structure of the artefact and, in most cases, are the dimensions that designers take into consideration (for example, in the case of the car, type of engine, type of suspensions, weight, etc.). Service characteristics, by contrast, are the 'services' actually delivered by the artifact in which users are interested (in the case of the car, speed, reliability, comfort, etc.). (Castaldi, Fontana and Nuvolari, p. 549).

Technical characteristics are the "means" (e.g., car engine) by which technological products exploit some inputs (e.g., fuel or electricity) in order to achieve given final outputs for each service characteristic (e.g., a given car speed in km/h). Various subcategories of service characteristics co-exist in each technological product: (1) main services, which determine in the first place the introduction of the technology; (2) complementary services, which facilitate the performance of main services; and (3) externalities, which are unwanted services jointly produced that must be minimized. All these types of service characteristics contribute, albeit with different weights, to define the total extent of effectiveness for customers of a technological product. This theoretical framework often has been adopted by technological change scholars. For instance, an empirical study on tank technology by Castaldi et al. (2009) shows that technical and service characteristics are very useful indicators to analyze the historical evolution of technological trajectories over time. Saviotti and Metcalfe (1984, p. 142) argue that the evolution of any technological product depends on:

- changes in the absolute values of technical characteristics and/or service characteristics;
- mixture or balance of one (or both) of these types of characteristics;
- changes in the relationships (the "pattern of mapping") between these types of characteristics.

The framework by Saviotti and Metcalfe provides three critical assumptions for the present analysis (summarized in Fig. 3.1). First, the final value of any

**Fig. 3.1** From technical characteristics to customer effectiveness (adapted from Schiavone 2013a)

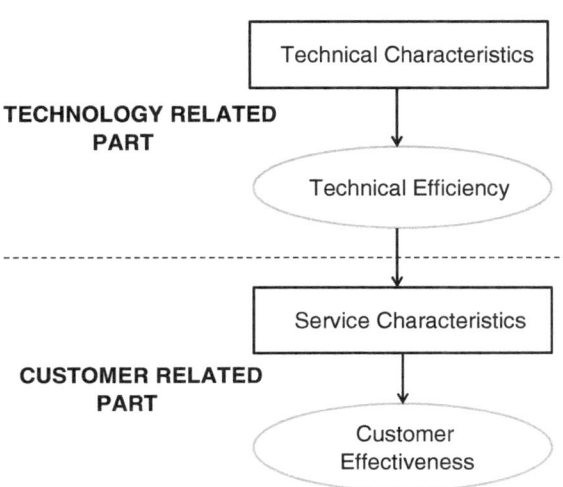

technological artifact depends on both its technology-related part (which is based on its technical characteristics and estimated in terms of technical efficiency) and its customer-related part (which is based on its service characteristics and estimated in terms of effectiveness for users). Both of them contribute to the processes by which technological products exploit some inputs and achieve given levels of performance (outputs). Second, technical efficiency depends directly on the technical characteristics of a product. Improvements (or changes in the set) in the technical features of an old product usually lead to improvements in its technical efficiency. Third, customer effectiveness depends directly on the service characteristics of technological products and indirectly on their technical characteristics. Indeed, improvements in efficiency of existing technical features lead to improvements in the levels of one or more service characteristics and, thus to a higher effectiveness of the product for customers. Also, the introduction of new technical characteristics may produce new service characteristics of old products and, thus, change positively their performance and customer effectiveness.

As they are traditional corporate approaches that do not seem to allow a further way, which instead Saviotti and Metcalfe (1984, p. 142) consider, to improve service characteristics and customer effectiveness of OTBPs: the achievement of these goals without any change or improvement in the technical characteristics of the OTBP. The revitalization of an OTBP improves its performances by an upgrade of its technical characteristics. For instance, an old formula 1 racing car has to improve, among others, the efficiency in the technical characteristic "design" (the more aerodynamic as possible) in order to improve the service characteristic "speed" and be more effective in its future version. In the case of retrofitting, home telephones have to be equipped with some new technical characteristics (e.g., display and keyboard of new generation) in order to be more efficient and perform the additional service characteristic "text messaging" increasing users' effectiveness.

Two main considerations can be developed about such hypothesized type of improvements. First, the assumed lack of changes in the technical characteristics of OTBP entails that improvements in its service characteristics must be achieved by networking the old product with another technological artifact. This new product has to work as an external means by which to improve the traditional OTBP performances. This artifact must be based on new and more efficient technology in order to improve the level of outputs of some service characteristics of the old one. These changes will lead, then, to improvements in effectiveness for customers. Second, improvements in performance of OTBPs not based on technical characteristics and efficiency but only on service characteristics do not produce more effectiveness for "all" customers (or members of a technology-based community of practice). In this case, the OTBP provides more effectiveness just for users (or community members) which, after technological change, evolved the set of their targeted needs in an "ambivalent way." On the one hand, these OTBP users need to upgrade a part of their needs and preferences in order to appreciate and exploit new technology. On the other hand, the same OTBP users, for some specific reason (e.g., high switching costs or personal habits, culture, passion, nostalgia, or professional needs), must "keep constant" a part of their traditional preferences and needs.

As discussed in the next subsection, converters and adapters can be useful technical means by which these considerations about this third way to improve the OTBPs' performance can occur in practice.

### 3.2.2 Compatibility Between Old and New Technologies

The improvement in performance of OTBPs without changes in their technical characteristics imply these artifacts must be compatible with other objects. Technological products are compatible when their design is coordinated in some way and are able to work together (Farrell and Saloner, 1986).[5] Technical compatibility can be established between compatible complements or compatible substitutes (David and Bunn 1988). In the first case, the implementation of a given process requires the integration of two different products (e.g., 35-mm film and camera). In the second case, users can use alternatively both the products to implement the process and achieve the same output (e.g., both Microsoft Word and Openoffice can be installed on PC and utilized to create the same types of documents).

Technical compatibility is critical to achieve economies of scale and demand-side benefits. These benefits are three (Farrell and Saloner 1986): (1) interchangeability of complementary products; (2) ease of communication between people and/or between machines; and (3) cost-saving. Technical compatibility is a critical firm-level factor affecting the characteristics of future industry dominant design. Three types of technical compatibility exist:

- horizontal compatibility: two technological systems, functionally equivalent, of the same generation are made compatible. Each format can gain by gateway devices the installed base and set of complementary goods of its competitor in order to become the new industry standard;
- backwards compatibility: two technologies of different generations are compatible as the new format is designed to be compatible with the old format. On the one hand, adapters allow new technology to get more easily the installed base of the previous format and achieve quickly critical mass (Ehrhardt 2004). On the other hand, gateway devices allow old technology to augment the amount of its complementary (new technology-based) goods;
- forwards compatibility: an old generation technology is already made forwards compatible with the new generation of itself (e.g., document formats—Microsoft Word 2003 is already compatible with future versions of Words). Forwards compatibility usually occurs without converters as the formats made compatible are just different versions (one older than the other) of the same product.

---

[5] Information technology scholars use the notion of interoperability in order to define the compatibility between different computers and/or software. Inter-operability, broadly speaking, is "the degree to which diverse systems, organizations, and/or individuals are able to work together to achieve a common goal" (Ide and Pustejovsky 2010).

The so-called adapter strategy (Shapiro and Varian 1998) is one basic mechanism bringing about backwards compatibility between new and old technological systems and products. Adapter strategy is also one of most valuable tactics for companies manufacturing technological formats losing "standard wars", in which different technologies compete to become the new industry standard. Adapter strategy improves (or at least supports) the network externalities of the old/losing technology by adding new complementary goods to its original technological system and plugging it into a larger/emerging network. Adapters can be very simple conversion technologies (e.g., adapters for electric devices in foreign countries) or even complex technological artifacts. In this second case, they are expensive to develop and risk working imperfectly (Katz and Shapiro 1994). Adapters bring about technological systems composed of incompatible technologies giving, generally, lower utility and performance than systems composed of fully compatible components (Farrell and Saloner 1992; Shapiro and Varian1998).

Different types of companies can implement adapters. On the one hand, a single firm acting unilaterally can manufacture an adapter in order to make compatible its products with those (usually dominant in the market) of another company. For instance, during the 1980s and 1990s Apple was used to develop adapters and converters to contrast the erosion of market shares of its programs against Microsoft software. On the other hand, adapters can be developed by third companies interested in capturing the benefits coming from the compatibility between different technological systems.

### 3.2.2.1   Emulation and Backward Compatibility

Emulation is an interesting solution for backwards compatibility by converters as it overcomes critical technical limitations of old technologies and products. In particular, gateway technologies emulating OTBPs can be widespread and very useful for old technology-based communities of users in which practices, skills, and expertise are basically more important than physical artifacts themselves.

Backwards compatibility is not always possible or easy to achieve. A number of technical problems might hamper the development of compatibility-based approaches. For instance, a typical issue is that the old product is not pre-arranged to be compatible with any future technology. In this case, converters and adapters can emulate the functions of specific technological systems in different settings. Emulation is a very common phenomenon in human beings and animal species. Charles Darwin in his "Descent of the Man and Selection in Relation to Sex" (1871) noted that birds sing for rivalry and emulation. In this view, emulation is a term synonymous of "imitation" by someone of a behavior usually adopted by someone else. In informatics an emulator is a "hardware or software that permits programs written for one computer to be run on another computer" (Merriam-Webster Dictionary). Overall, emulation in this field of technology can be defined as the creation of an artificial environment, within a real environment, allowing actions and behaviors originally though for another real environment.

Graphics provides an interesting example of how emulation supports customers' effectiveness and the survival of their  knowledge and expertise related to the use of a technological product becoming obsolete after technological change. Until a few decades ago, designers were used to draw with just stylus and sheets. Within the last 30 years, technology substitution and change for the implementation of this practice have been radical. Software (e.g., AutoCAD or Photoshop) and computers equipped with powerful video cards became the most important technological devices for many architects, fashion designers, and other professionals. Their practices became based on "Computer Aided Design" (CAD) systems. The core of technological change toward CAD systems refers to the interface used by designer: traditional designers were used to utilize stylus, whereas CAD professionals draw with PC mouse. New interface and technology for this practice have many obvious advantages for professional designers but also some problems. The main refers to the domain of ergonomics: designers' sensitivity and manual ability with stylus are not anymore critical competencies in order to implement this practice with CAD systems. The traditional manual "art" and expertise in handling a stylus is lost with the new interface. This is the main reason whereby some professional designers use graphics tablets. In general, a tablet is any computer device equipped with a touchscreen or stylus (e.g., booklets, slates, and convertible notebooks). A basic graphic tablet consists of two components: (1) a stylus by which designers can draw via their own hands and (2) an electronic tablet which receives the inputs of the stylus, converts them into digital form, and transmits these inputs to PC. Therefore, the designs traced on tablet are reported on PC monitor, if necessary they are manipulated via mouse and software and, afterwards, are stored as image files. Tablets make compatible the usual technological interface of designers (stylus) and some new artifacts (PC, software) of CAD and digital technologies. The key benefits for design and graphics practices offered by these devices is the digitization of the inputs given by the user (the designer).

Graphic tablets emulate the outputs of traditional handcraft practice of designers into a digital-based environment. Graphics tablet is a useful conversion device for professional skilled designers for ergonomic and cognitive reasons related to their old technological interface. It preserves their manual sensitivity and experience in the use of stylus from a radical replacement due to the emergence of as innovative interface (PC mouse). The risk of arm injuries due to the prolonged use of mouse is another critical reason supporting the diffusion of this device within the community. Indeed, the use of stylus is much more natural and less repetitive than the use of mouse.

A common problem with emulation is the risk of legal issues that sometimes might emerge (Shapiro and Varian 1998). For instance, international laws often forbid unofficial videogames emulators by protecting the intellectual property rights of the software companies that developed the emulated videogames.

## 3.3  Vintage Innovation

Drawing on the Saviotti and Metcalfe framework and the assumptions on converters and adapters, this section outlines a third approach to improve the service characteristics and customer effectiveness of OTBPs. Vintage innovation is the name proposed for this approach (Schiavone 2013a). The term "vintage" is used since this approach is likely to be suitable especially for the improvement of customer effectiveness of those OTBPs receiving the enduring interest of some groups of users even after the emergence of new competing technologies (and their related products) offering superior performances. The term "innovation" is used because this approach entails the development of a new product based on the new competing technology.

The phenomenon of vintage innovation is a rising strategic reaction to technological change for (and by) communities of practice focused around an old technological artifact. Indeed, this approach can be promoted by companies or even end-users of OTBPs. The fact that this approach can be performed also by end-users (as shown in Chap. 5) suggests vintage innovation should be categorized, more generically, as a "phenomenon" rather than as a corporate approach.

The aforementioned theoretical considerations suggest that vintage innovation has two critical characteristics:

1. This approach improves directly the customer effectiveness of OTBPs without modification of their technical characteristics. Vintage innovation, thus, entails the development of a new-technology-based product (a conversion device) aimed at improving the "customer-related" part of the OTBP.
2. Vintage innovation improves customer effectiveness by producing improvements in the traditional outputs in some of (but not all) the service characteristics of the old product. Indeed, this approach is based on the assumption that after technological change the set of needs of members of communities centered around OTBP is constant in relation to some service characteristics and changes for others.

Compatibility between different technological systems is crucial in the present argument. In vintage innovation, communication and interoperability between different technological paradigms occur through the development of new technology-based "bridging products," here named vintage products. Technically speaking, vintage products are converters establishing a technical compatibility between different platforms in order to meet customer needs. These new products are developed in order to be technically (backwards) compatible and interoperable with existing ones and, consequently, to improve the normal performances of the latter with additional service characteristics (commonly, the acquisition and elaboration of data from other technological devices). However, the old technological product does not change its technical structure.

This argument is strongly related to the principle of modularity. This concept refers to the division of a product or process into parts (called modules), which can then communicate with one another only through standardized interfaces within a standardized architecture (Baldwin and Clark 1997). Such architectures shape, at any unit of analysis (e.g., final products, product subcomponents), modular systems (Schilling 2000). Interoperability and compatibility can be established just if OTBPs are already prearranged to be interoperable as units of a modular system, made by more combinable components (namely, other OTBPs). In other words, interoperability between old and new technologies by vintage innovation grounds on previous interoperability between existing technological devices. If modularity is not technically feasible, then a possible solution might be to implement vintage innovation by the emulation of the old product in innovative environments and fields of application.

Vintage products provide an interface of communication between old and new technologies. By this interface, an OTBP can exchange data and information with the new technological system and improve its performance and customer effectiveness. Using Ronald Burt (1992) words, vintage innovation fills in a structural hole[6] between formerly competing and unconnected technologies and products.

The technological architecture of vintage products is quite complex. These products integrate three different modules of technological knowledge in order to improve service characteristics and customer effectiveness of OTBPs:

1. knowledge about the old declining technology;
2. knowledge about the new competing technology;
3. knowledge about other technologies (e.g., informatics or electronics) necessary to allow an effective exchange of data and information.

Vintage products ground on the theoretical assumption that new technology and complementary technologies offer great opportunities to improve OTBPs performances and effectiveness (Saviotti and Metcalfe 1984; Prencipe 2003).

Vintage innovation provides benefits for both the OTBP and new competing technology: (1) the former does not undertake, by revitalization, an unsuccessful performance-based race with the latter; (2) new technology experiences an easier entry of new technology in the segments of OTBP's loyal users.

The remainder of this section reports the main characteristics of this emerging approach and its critical differences with the traditional approaches for the improvement of performance of old declining products.

## 3.3.1 Differences with Similar Approaches

Vintage innovation by firms extends the traditional categorization of the corporate approaches by which improvements in technical performance and effectiveness for customers of OTBPs occur (Fig. 3.2).

There are various key differences between vintage innovation and revitalization/retrofitting, approaches based on improvements of technical efficiency. The first difference is at the technological level. Drawing on the assumptions of the framework by Saviotti and Metcalfe (1984), vintage innovation provides a scope of improvement narrower than other approaches. Vintage innovation focuses only on

---

[6] A structural hole is "the separation between non redundant contacts. Non redundant contacts are connected by a structural hole. A structural hole is a relationship of non-redundancy between two contacts" (Burt 1992, p. 18).

**Fig. 3.2** Approaches for the improvement in performance of old technological products (author's elaboration)

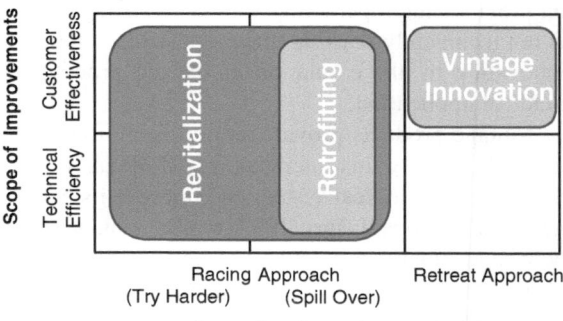

service characteristics of the OTBP in order to improve its customer effectiveness. Instead, standard revitalization and its variant retrofitting focus on both technical characteristics (in order to improve technical efficiency) and service characteristics (in order to improve customer effectiveness) of the old product.

In standard revitalization and retrofitting OTBP improvements are necessarily coupled with an R&D activity to innovate the old product. Instead, OTBP improvements and R&D efforts of firms are de-coupled in vintage innovation as this approach does not change, for instance, the external form of an OTBP (as standard revitalization) or integrate it with the new emerging technology into a unique hybrid technological artifact (as retrofitting).

Vintage innovation differs from both the traditional efficiency-based approaches also for the nature of the relationship which ties together OTBPs and the new competing technology. In vintage innovation the improvement of OTBPs is not the outcome of technology competition (as in revitalization). In vintage innovation, old and new technology collaborate (as retrofitting) but in different ways. Indeed, in this approach, OTBPs work together and become compatible with new technologies and their related products but by the intermediation of an external converter. However, the OTBP remains discrete and does not merge with new technology into a unique artifact (as retrofitted cars or innovative home telephones).

The third relevant difference between this new approach and traditional racing approaches based on revitalization is at strategic level. The main goal of standard revitalization and retrofitting is to contrast the entry of the new technology. These reactions (as shown by the phenomenon of sailing ship effect) extend the lifetime of the OTBP also by its transition from mass market to market niches. Revitalization and retrofitting are both racing approaches as they improve the technical efficiency and performances of the main services of OTBPs. Revitalization is based on a "try harder" strategic response of old technology to new technology. Retrofitting is based on the "spill over" and fusion of the new technology on the old (Adner and Snow 2010).

The key purpose of vintage innovation, instead, is to support the entry of new technology within the market niches in which customers still use the OTBP. Therefore, vintage innovation should be in theory a retreat reaction. However, in practice, this approach is an intermediate solution between these two opposite behaviors. Indeed, vintage innovation also contrasts and/or postpones technological substitution by improving the services characteristics of the old product (and consequently the customer effectiveness).

Another key difference refers to the OTBP life cycle. Revitalization usually occurs when old technology starts its decline and experiences its performance limits and the OTBP starts becoming a niche product (Howells 2002; Schiavone 2011). Instead, vintage innovation fits better as the "very last" step on the evolution path of the OTBP from a mass product to a niche product. This approach should occur when the "race" of old technology and its product are definitively lost and new technology already affects the set of needs, capabilities, and desired performances of OTBP users.

## 3.3.2 Vintage Innovation and Customers

On the technology-side, modularity of OTBPs and backwards compatibility are some of the main technological variables that firms must take into account in order to promote vintage innovation. However, the proper and successful implementation of vintage innovation depends on various conditions related to OTBPs' customers, ideally forming old technology-based communities of practice.

A critical precondition for the success of this approach is that customers, although they continue to prefer old technology, are able to use the new technology and recognize it, can improve in some way the standard performance and effectiveness of their old products becoming obsolete. Although this is a general closeness toward technological substitution within the CoP, vintage innovation has to give CoP members the opportunity to benefit partially from the progress and advantages of the new rising technology. Backwards compatibility offers the new technology to the community and, to some extent, supports the utilization, recombination, and diffusion of bundles of new technological knowledge and practices between its members. This approach is also likely to make easier and more affordable even the entry of new members into the CoP (as shown in Chap. 4 by the case of analog photographers community and lomographers).

If OTBP users satisfy this precondition, then vintage innovation can take place. Of course, not every old technological product used by these communities will be a suitable target for this innovative approach. A critical condition for the implementation of vintage innovation is that users hold substantial prior experience, skills, and technological knowledge (achieved through their former use) about the OTBP. Vintage products give to CoP members the possibility to continue using these intangible communal resources even after technological change. This condition gives users a twofold advantage in terms of switching costs as people "recover" their own prior technological expertise and do not feel lost the time and efforts they spent in learning how to use the old product. Prior knowledge survival is quite important especially in those communities in which hierarchies between members are mainly based on the proper use of a given artifact. For instance, the skills in playing musical instruments outline the leader members of a local orchestra. Therefore, vintage innovation is an effective approach solely for a community old product, which requires a minimum of knowledge or expertise to be used. For instance, floppy disks are not good for vintage innovation as their use does not require any type of expertise or knowledge. Indeed, users were not disappointed to switch to newer and more advanced substitutive products.

Finally, users must give a recognized symbolic meaning and high social relevance to the OTBP, which therefore should be a community socio-technical ensemble (Pinch and Bjiker 1984) with specific social resources and structures (Orlikowski 2000). This precondition allows vintage innovation to satisfy a relevant need for OTBP users after technological change: to avoid the risk that, in the long term, some external changes (e.g., the discovery and introduction in the community of new ways of implementing its core practices with new technology) might reshuffle or weaken the CoP original identity and structures based on their core artifact becoming obsolete.

Vintage innovation leaves unaltered these dimensions so that users can be still basically focused around their old core product.

## 3.4 Vintage Innovation and Corporate Strategies

Despite vintage innovation being a phenomenon that not only companies but also customers of OTBP can implement, the last section of this chapter analyzes the main strategic implications of this approach for industrial and service organizations. Indeed, vintage innovation can affect the decisions and strategies of both incumbent and emerging companies at least in three strategic domains: technology management, innovation strategy, and marketing.

### 3.4.1 Technology Management

Vintage innovation outlines a number of new challenges, evaluations, and implications for companies in the field of technology management. The main technological challenge for firms developing this approach is to establish modularity and collaborative relationships (even by emulation) between the vintage products, old products, and new technology (and related products).

Backwards compatibility provides new technology-based inputs improving the traditional levels of outputs of the OTBP service characteristics. Compatibility and interoperability imply the OTBP must already belong to some modular technological system. Modularity is critical in the context of vintage innovation as this system property sets out the extent of the possible forms of backwards compatibility between the OTBP and new technology. This evidence is interesting as it shows that modularity at system level (the old product functioning depends on a number of complementary goods) is another further avenue of improvements for OTBPs after technological change.

Backwards compatibility, at least theoretically, gives to old technology and its products a "second life." The connection of an old technology with its external fast-changing technological environment (namely, complementary technologies) is a critical issue in order to support its survival over time. In vintage innovation, the old technology S-curve extends mainly horizontally (and grow slightly on the vertical axis) on the Foster's model (1986), since its performance will not improve significantly (but it will improve). The temporal presence within some portions of the market is, instead, the main dimension of the old technology survival by vintage innovation.

An important variable to consider in this domain for the implementation (or not) of vintage innovation is the pace of emergence and success of new technology. If technological change is fast and disruptive, then the decline phase of OTBPs will be shorter than other old products for which the introduction of new technology was not disruptive (e.g., cars). In this case, new technology soon starts improving its performance and backwards compatibility soon becomes really effective also for old technology users. Moreover, a short decline phase of the OTBP makes it easier for customers to keep alive and not "forget" their old technology-based practices. Conversely, a "life after death" for the OTBP is less necessary if the transition from old to new technology is slow and OTBP customers have enough time to get familiar with (and redefine their practices by) new technology.

Vintage innovation is not a disruptive approach[7] since it is based on the development of technological innovations (the vintage product) for niches and small segments of the old technology market. Therefore, vintage innovation provides various advantages to old technology companies: (1) the exploitation of some of the R&D competencies and knowledge related to the old technology; (2) the exploitation of market knowledge and/or brand awareness achieved when the old technology was dominant; (3) the possibility to achieve a dominant position and a critical competitive advantage in the market niches resisting the new technology and with "ambivalent" set of needs by users. However, vintage innovation is a technology management option also for new companies with strong resources, capabilities, and competencies about the emerging technology.

An in-depth knowledge of CoPs technical skills, habits, and culture is another critical requirement in order to understand the most important CoP old products

---

[7] Christensen (1997) distinguishes between disruptive innovation and sustaining innovation. A disruptive innovation is a new technology which has the potential to revolutionize an industry and, for this reason, many companies tend to ignore it. This type of innovation therefore is similar to the notion of radical innovation (Henderson and Clark 1990). Instead, a sustaining innovation does not create new markets or value but only evolves existing ones with better value. In this sense it is similar to the notion of incremental innovation (Henderson and Clark 1990).

deserving the support of a vintage innovation. The centrality of the technological side of the community for vintage innovation explains why sometimes this phenomenon in some cases (e.g., communities in which members exchange and share skilled technological capabilities and engineering knowledge) is promoted directly by the OTBP users (as reported in Chap. 5).

### 3.4.2 Innovation Strategy

The essence of this demand-pulled innovation approach is that firm develops and launches a vintage product establishing backwards compatibility between new technology and an OTBP. To this end, firms implement an innovation strategy here named "technological reverse" (Fig. 3.3). Indeed, companies can find critical guidelines for new product development not just by working on the emerging technological paradigm but even by looking "back" to its technological "predecessor." This innovation strategy, having its core activity in the technological development of vintage products, should be a part of a larger innovation strategic design in order to be really powerful. Due to its high specificity, the best application of technology reverse is probably in combination with other innovation strategic behaviors. .

This formula of innovation makes partially misleading the common distinction between offensive (pioneers) and defensive (followers) innovation strategies. Indeed, firms innovating through vintage products are both pioneers, in seeking

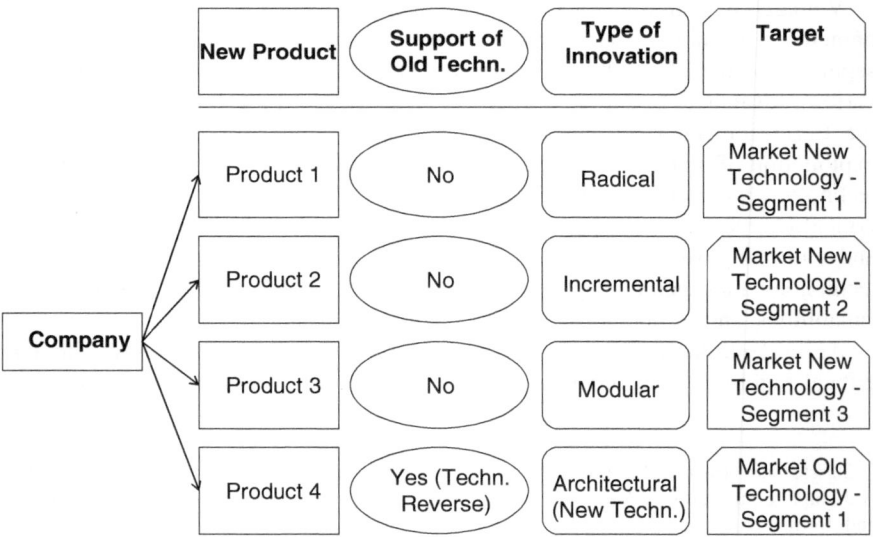

**Fig. 3.3** A hypothetical innovations portfolio of an incumbent firm implementing technology reverse after technological change (adapted from Schiavone 2013b)

new business opportunities (not yet exploited by competitors) related to old technologies, and followers, as they do not search for technological breakthroughs. So, these organizations must be considered as hybrid innovators.

The implementation of this approach may have two alternative meanings in terms of technological strategy of incumbent firm. First, if the firm was already serving the "vintage niche" before technological change, then vintage innovation is likely to be a key step in the consolidation strategy of the company within its usual market segment. Second, if the incumbent firm retrenched to the "vintage niche" after technological change, then the implementation of vintage innovation could demonstrate the inability of the organization to face technological change. Technology reverse can be implemented by both companies formerly producing a dominant design and/or selling its complementary products.

The development of complex and sophisticated artifacts as vintage products implies critical managerial capabilities. Managers have to face more strategic, operational, and organizational issues typical of "participation strategy" (Cooper and Smith 1992) by which companies develop and sell products and services for both new technology market and old technology market.

The main strategic limit of this approach for innovation strategy is that it cannot be successfully applied to old ordinary products, simple to use, and without any community of loyal users. The implementation of technology reverse to "resuscitate" these products would be superfluous and, probably, not appreciated by the market.

## 3.4.3  Marketing

The third relevant domain for vintage innovation is marketing. In general, this approach is in line with the so-called phenomenon of retro-revolution (Brown 2001), revival of old brands and traditional images in consumers' minds, and companies' marketing strategies. Retro-marketing strategies have an appeal just for some particular customers (e.g., nostalgic users of old outdated products). Similarly, vintage innovation is the key element of niche marketing strategies aimed at giving added value not to all the market but just to some specific niches of OTBP customers with similar characteristics.

Potential users of vintage products should balance harmonically the trade-off between resistance to change and technological innovation. As already discussed in Chap. 2, this is a typical characteristic of old technology-based communities of practice. Therefore, every marketing action targeted for customers of vintage products should take into account that these users have both static and dynamic needs in relation to the service characteristics of their beloved OTBP. Static needs are not affected by technological change (users do not search for better performance), whereas dynamic needs evolve and become more demanding with technical progress.

Marketing strategy for vintage products must be based and take advantage of such needs' ambivalence typical of these retro-communities.

The well-known 4P model developed by Jerome McCarthy (1960) and widespread by Philip Kotler (1988) over many decades is a useful framework to outline how the marketing mix of a vintage product could be:

- Product: the assumption that users have strong technological skills supports the development of sophisticated technical artifacts, based on both hardware and software components. Moreover, vintage innovation can be a component of a larger marketing strategy based on the development of more artifacts and/or services for old technology-based communities.
- Price: this approach offers many possibilities to companies to carry out differentiation strategy. The specific services provided by vintage products are likely to be sources of premium price.
- Placement: the fact that companies know ex-ante who are their potential customers supports the adoption of direct channels of selling. Both traditional and digital channels can be useful to this end.
- Promotion: unless the community of users is large enough, traditional advertising for vintage innovation is not necessary. Communication and promotion by digital technologies are powerful tools to achieve this end. Therefore, companies should provide customers technological infrastructures to communicate and promote vintage products.

In sum, the success of vintage innovation should be based on both the satisfaction of the static and dynamic needs of the OTBP users and a proper arrangement of the marketing mix of vintage products.

# References

Adner R, Snow D (2010) Old technology responses to new technology threats: demand heterogeneity and technology retreats. Ind Corp Change 19(5):1655–1675

Baldwin CY, Clark KB (1997) Managing in the age of modularity. Harvard Bus Rev 75(5):84–93

Brown S (2001) Marketing: the retro revolution. SAGE publications, Thousand Oaks

Burt RS (1992). Structural holes: the social structure of competition. Harvard University Press, Cambridge

Castaldi C, Fontana R, Nuvolari A (2009) Chariots of fire: the evolution of tank technology, 1915–1945. J Evol Econ 19:545–566

Christensen C (1997) The innovator's dilemma. When new technologies cause great firms to fail. Harvard Business School Press, MA

Cooper AC, Smith CG (1992) How established firms respond to threatening technologies. Acad Manag Executive 6(2):56–69

David PA, Bunn JA (1988) The economics of gateway technologies and network evolution: lessons from electricity supply history. Inf Econ Policy 3:165–202

Drucker PF (1985). Innovation and entrepreneurship. HarperCollins, New York

Edgerton D (2006) The shock of the old: technology and global history since 1900. Oxford University Press, New York

Ehrhardt M (2004) Network effects, standardisation and competitive strategy: how companies influence the emergence of dominant designs. Int J Technol Manage 27:272–294

Farrell J, Saloner G (1986) Installed base and compatibility: innovation, product preannouncements, and predation. Am Econ Rev 76:940–955

Farrell J, Saloner G (1992) Converters, compatibility, and the control of interfaces. J Ind Econ 40:9–36

Foster RN (1986) Innovation: the attacker's ADVANTAGE. Summit Books, New York

Fukuyama F (1992) The end of history and the last man. Avon Books, New York

Howells J (2002) The response of old technology incumbents to technological competition: does the sailing ship effect exist? J Manage Stud 39(7):887–906

Ide N, Pustejovsky J (2010) What does interoperability mean, anyway? Toward an operational definition of interoperability for language technology. In Proceedings of the 2nd International Conference on Global Interoperability for Language Resources (ICGL) New York

Katz M, Shapiro C (1994) Systems competition and network effects. J Econ Perspect 8:93–115

Kojève A (1980). Introduction to the reading of Hegel. Cornell University Press, New York

Kotler P (1988) Marketing management: analysis, planning, implementation and control, 6th edn. Prentice-Hall, NJ

Madhavan R, Grover R (1998) From embedded knowledge to embodied knowledge: new product development as knowledge management. J Mark 62(4):1–12

McCarthy JE (1960) Basic marketing. A managerial approach. Richard D. Irwin, IL

Orlikowski WJ (2000). Using technology and constituting structures: a practice lens for studying technology in organizations. Organ Sci 404–428

Pinch TJ, Bijker WE (1984). The social construction of facts and artefacts: or how the sociology of science and the sociology of technology might benefit each other. Soc Stud of Sci 399–441

Prencipe A (2003) Corporate strategy and systems integration capabilities: managing networks in complex systems industries. In Prencipe A, Davies A, Hobday M (eds) The business of systems integration, Oxford University Press, Oxford

Saviotti P, Metcalfe JS (1984) A theoretical approach to the construction of technological output indicators. Res Policy 13(3):141–151

Schiavone F (2011) Strategic reactions to technology competition: a decision-making model. Manag Decis 49(5):801–809

Schiavone F (2013a). Vintage innovation: how to improve the service characteristics and customer effectiveness of products becoming obsolete. IEEE Trans Eng Manage 60(2)

Schiavone F (2013b). Innovation approaches for old products revitalisation after technological change: the rise of technology reverse. Forthcoming Int J Innov Manage

Schilling MA (2000) Toward a general modular systems theory and its application to interfirm product modularity. Acad Manag Rev 25(2):312–334

Shapiro C, Varian HR (1998) Information rules: a strategic guide to the network economy. Harvard Business School Press, MA

Vico G (1725). The new science of Giambattista Vico. Reprinted in 1984. Cornell University Press, New York

# Chapter 4
# Vintage Innovation by Firms

**Abstract** This chapter reports two case studies (vinyl emulator for disk jockeys and film scanners for analog photographers) illustrating the corporate approach of vintage innovation. The common characteristic of these case studies is that vintage innovation is developed by companies. The unit of analysis of the study is the improvement, by vintage innovation, of customer effectiveness of old technological products after the emergence of new competing technologies. Three different dimensions are illustrated for each case in order to provide a comprehensive understanding of the phenomenon: (1) The process of technological change; (2) The community of practice centered around the OTBP; (3) The vintage innovation. The study is based on primary and secondary data. The chapter ends with the main managerial implications of vintage innovation for companies.

**Keywords** Improvement of performance • Dj turntablists • Vinyl emulator • Analog photography • Film scanners • Lomography

## 4.1 Introduction

This chapter reports two case studies (vinyl emulator for disk jockeys and film scanners for analog photographers) illustrating the corporate approach of vintage innovation. The common characteristic of these case studies is that vintage innovation is developed by companies.

Yin (1994) suggests that case study research method is suitable when the form of the research question is "how" or "why," there is no need for control on events and the analysis is about contemporary events. The key rationale of an illustrative

F. Schiavone, *Communities of Practice and Vintage Innovation*,
SpringerBriefs in Business, DOI: 10.1007/978-3-319-01902-4_4,
© The Author(s) 2014

case study is the development of a "Weberian ideal-type"[1] already existing in practice. Illustrative case studies achieve this goal, since "these are case studies which provide empirical exemplars of the embodiment of particular theories… they provide illustrations of the way in which particular theoretical categorizations can be observed in practice" (Scapens 2004, p. 259). Articles reporting new theory about technological change often use the method of illustrative case studies (e.g., Faulkner and Runde 2009). An information-oriented selection led to the analysis of vinyl emulator and film scanner, clear cases of vintage products aimed at improving the customer effectiveness of old technological products. This type of case selection was chosen since extreme/deviant cases are useful "to understand the limits of existing theories and to develop new concepts, variables and theories that are able to account for deviant cases" (Flyvbjerg 2011).

The unit of analysis of the study is the improvement, by vintage innovation, of customer effectiveness of old technological products after the emergence of new competing technologies. However, three different dimensions are illustrated for each case in order to provide a comprehensive understanding of the phenomenon:

- the process of technological change;
- the community of practice centred around the OTBP;
- the vintage innovation.

The study is based on primary and secondary data. Primary data were collected for the development of the second case study by a number of interviews, conducted by electronic mail, to small samples of analog photographers. The respondents are individuals from various nations (e.g., USA, Canada, Germany, Australia) and often are members of virtual CoPs. At the same time, a desk research was implemented to collect secondary data and information for the study. This technique was adopted because desk research "is an excellent tool for putting together pictures of a marketing environment—showing the market size, the suppliers, the products that they make and the trends in the market" (Hague et al. 2004, p. 47). Both the cases report secondary data collected mainly from two types of sources of data: (1) Internet websites (e.g., corporate websites, users' forums); (2) scientific references.

These methodological prescriptions are taken into account also for the treatment of the two case studies (multimedia software for radio-amateurs and new generation software for arcade videogame players), reported in Chap 5, about vintage innovation developed by communities of OTBP users. Therefore, the present research overall reports a total of four illustrative cases studies, which is considered the minimum amount of empirical evidences necessary for the definition of new theoretical constructs (Eisenhardt 1989).

---

[1] "An ideal-type is formed by the one-sided accentuation of one or more points of view and by the synthesis of a great many diffuse, discrete, more or less present and occasionally absent concrete individual phenomena, which are arranged according to those one-sidedly emphasised viewpoints into a unified analytical construct" (Weber 1949).

## 4.2  Vinyl Emulator for Disk Jockeys[2]

### 4.2.1  Analog Music, Turntables, and Technological Change

The recent history of music is rich with interesting examples for scholars of technological change and innovation. Over the last 20 years, technological change in the music industry mainly related to the shift from analog sound toward digital sound. Drawing on the set of notions developed by Dosi (1982), analog sound and digital sound may be considered as two different technological paradigms satisfying the same problem (or users' need) within the music industry. Over time more technological trajectories emerged within each paradigm: vinyl and audio-cassette in the analog paradigm; compact disks and mp3 in the digital paradigm. In this industry, technological progress over time may be referred, for instance, to the trade-off between the quality of sound and the price of audio supports.

Within the "vinyl trajectory" of analog paradigm turntables were the dominant product for a specific type of end-users; disk jockeys (henceforth: DJs). The business of technologies for DJs is a quite recent sub-industry within the larger music industry. The first technological companies operating in this industry segment appeared in the middle of the last century (e.g., Stanton was founded in 1946 and Technics, division of Panasonic, in 1969). The number of these companies grew quickly over time due to the parallel development in Western countries of some related industries (e.g., radio industry or night entertainment industry) which increased the demand for DJs services and the need for related technologies. This sub-industry segment is technologically diversified and complex as its products merge more technological fields (e.g., physics, electronics, sound engineering, informatics).

Before technological change, DJs were used to utilize their turntables jointly with many other technological products to play records and carry out their performances: cartridges, mixer, headphone, microphone, and pre-amplifiers. Typical capabilities and skills of expert DJs are beatmatching, mixing, and scratching of songs recorded on vinyl (Brewster and Broughton 2000). From the viewpoint of a DJ, the main service characteristics of a turntable are "Vinyl Playing" and "Disk Jockeying." Indeed, these services show the level of implementation of the main DJs capabilities and practices, as scratching.

Both these services produce effectiveness for DJs thanks to two complementary services: (1) the outgoing of audio signals toward external technological devices (e.g., mixers or pre-amplifiers), (2) beat adjustment of the vinyl speed (for beatmatching and mixing two songs). Two main externalities characterize turntables: (1) the instability (vibration) of turntable, for instance while DJs scratch vinyl or due to the "skating" effect of the turntables tonearm; (2) damages to vinyl due to frequent use in the long term.

---

[2] This section is based on and extends some elements of the analysis developed by the author in Schiavone (2013a).

**Table 4.1**  Technical characteristics and service characteristic of an analog turntable

| Service characteristics | Technical characteristics |
|---|---|
| Vinyl playing and Djing (Main services) | Drive system |
| Audio outgoing and beat adjustment (Complementary services) | Pick-up system Arm system |
| Vinyl damage and instability (Externalities) | Speed control |

The manual of Technics SL-1200 is a valuable source of information in order to outline the key technical characteristics of a typical analog turntable used by DJs: (1) the drive system, producing torque and the rotation of turntable; (2) the "arm sub-system," composed of several technical components (e.g., arm-height control, arm lock, arm rest, arm clamp, and anti-skating control); (3) the "pick-up system," aimed at extracting the sound from vinyl by the application of another product on the turntable (magnetic cartridges); (4) the control of the turntable's speed of rotation (by a "pitch," a device used by DJs in order to mix properly the songs playing on two different turntables). Table 4.1 reports a possible matrix of technical and service characteristic of an analog turntable.

When the digital paradigm emerged, CD players quickly became dominant digital products as audio-cassette and turntables were over the analog paradigm. CD sales surpassed vinyl sales in 1988, just 5 years later the first commercialization of this digital device in the mass market (van den Dobbelsteen 2008). The rationale of such fast and disruptive substitution process is that digital sound is technologically more advanced than analog sound and offers a better trade-off between sound quality and the audio supports' price.

Referring to the market segment of DJs, in a few years a special type of CD players for DJs (compact disk jockeys, henceforth: CDJs) emerged in order to substitute traditional analog turntables (whose dominant design was the Technics SL-1200). Pioneer Electronics produced and launched on the market the first CDJ in October 1994 (CDJ is even the name of this Pioneer products line). CDJs are CD players with additional service and technical characteristics expressly designed with DJs in mind. CDJs have a pitch controller like an old turntable. But a CDJ usually has many additional technical and service characteristics that analog turntables cannot hold: BPM (beats per minute) counter, master tempo, countdown, cue button, and jog wheel. In particular, this last new technical feature is very useful for DJs. Jog wheel gives to DJs the capability of touching "virtually" the CD and moving it forward or backward as he/she could do with a traditional vinyl playing on a turntable.

A critical advantage of CDJ versus turntables is that it plays digital music recorded on CDs. Digital music is much cheaper and easy to find (e.g., via Internet) than analog music recorded on vinyl. Also, widespread PC hardware and music software, critical complementary technologies for DJs utilising CDJs, supported the technological change within this segment over time.

The emergence of CDJs started dissolving the dominant position of turntables within the DJ market segment. Since the first years of this century, a relevant share of DJs in the world began utilizing CDJs and "pensioned" their traditional turntables. However, such technological substitution did not occur homogeneously within the DJs community. Indeed, the value of analog turntables and the ability to use "hands" on vinyl are still considered critical conditions to DJs turntablists, a particular sub-community within the larger DJs community.

## 4.2.2   The Community of DJs Turntablists

The words "disk jockey" were used for the first time on the American magazine "Variety" in 1941 and, in general, they refer to someone presenting and playing music into a radio channel or someone playing records into a discotheque (Brewster and Broughton 2000).

For many years, analog turntables were the most common and widespread technological product in the sub-industry of equipment for DJs. The typical DJs turntable plays vinyl, mainly 33 and 45 "revolutions per minute" (RPM). The centrality of this product reinforced the development of a "turntables culture" within DJs community worldwide and over time (Souvignier 2003). Many firms manufacture and sell worldwide several types of DJs turntables: Stanton, Gemini, Vestax, Numark. However, Technics SL-1200 (and 1210) have always been (and, currently, still are) the dominant design of DJs turntables. SL-1200 is a Panasonic turntable produced since 1972 and was still manufactured until 2010.

DJs turntablists are a special type of turntable users as they form a proper community of practice. Turnablism is the "art of creating music via turntables" (Smith 2000). The term "was born out of their [turntablists] devotion to hours of practice and their strong sense of community. Many of these turntablists will insist on only using the Technics SL-1200 series turntables" (Lippit 2006). Such practice implies that the audio signals outgoing from turntables to external devices are of two types:

- the original audio information recorded on vinyl;
- distinctive sounds distorted by DJs commands with hands (scratching).

A number of evidences outline the symbolic and social centrality of turntables within this sub-community. For instance, the ability in using turntables is critical for young DJs and rappers in order to receive the appreciation of the other members of their community, and in time, to become themselves respected community members (Smith 2007). Moreover, the DMC (Disco Mix Club) DJ World Championship, the oldest and most important international contest for DJs turntablists (the first edition was in 1985), is one example of how the traditional technological practices and skills of turntablists are both revered and celebrated within this community. Indeed, participants to this international contest still use analog turntables for their performances.

In sum, the proper use of traditional turntables (the core practice of this subcommunity) is still crucial for DJ turntablists and this need minimizes the risk of technological substitution of the OTBP with CDJs. However, digital music provides at least two interesting possibilities for the improvement of DJs turntablists' core practices, performance, and effectiveness. First, music in digital format is easier to buy (e.g., by downloading from Internet) and to transport than analog vinyl. Second, the use of PC software and other digital devices support DJs in their performance for music creation (e.g., special effects, recording of tracks).

DJs are aware of these benefits and, despite the community being resistant to substitute analog turntables as its core technological artifact, turntablists are basically open to the adoption of digital devices. Consistently with this hybrid behavior, Faulkner and Runde (2009, p. 452) note that "… although some virtuous turntablists like Richard Quitevis (DJ Q-bert) and Ronald Keys (DJ Swamp) have embraced electronic tools (e.g., in the use of computers, drum machines, samplers, and so on), they retain their traditional turntable setups."

### 4.2.3 Vinyl Emulator

To improve customers' effectiveness for DJ turntablists, many companies commercialising DJ equipment (e.g., Rane, Stanton, Numark and Torq) have launched PC vinyl emulators over the last decade. The main function of this converter is to improve the technical performance and customer effectiveness of declining analog turntables without changing the interaction between this OTBP and users. The following elements compose this vintage product (Fig. 4.1):

- a special mixer;
- software for a PC;
- vinyl disk touching surface hardware.

Vinyl emulators improve the traditional set of service characteristics of analog turntables by connecting them to some complementary devices of CDJs, the new technological product directly competing with traditional turntables. DJ turntablists adopting vinyl emulator use traditional turntables as interfaces to send digital music commands to a PC. First, the DJ chooses which song to play. Then, he/she assigns, via the vinyl emulator software, the digital file corresponding to that song (and recorded on CD or stored in PC hard-disk) to the turntable. At this point, the format vinyl, playing on the turntable, "incorporates" the selected digital file coming from the PC hard-disk. Therefore, vinyl emulator allows to DJs to touch and "jock" with a traditional (analog) vinyl that temporarily plays a song not recorded on it but coming from an external (digital) source.

From customer perspective, this vintage product improves the OTBP performance by establishing backwards compatibility between analog turntables and CDJ complementary products (PC, mp3 files, software) and networking their competing technological paradigms (analog sound and digital sound paradigms). Turntables are just connected (and not fused) with new technological audio-digital devices. Thus, no spill-over of the new technology on the OTBP occurs such as no changes in the turntables' technical characteristics. The impact of vinyl emulator on the set of service characteristics of analog turntables is threefold.

- The vintage product improves the complementary service characteristic "audio outgoing" by a deep change of its inputs and outputs. Turntablists, indeed, now implement their scratching activity by substituting music from vinyl with music extracted from additional sources (digital files). Turntablists keep using analog

**Fig. 4.1**   The vinyl emulator (adapted from www.rane.com)

turntables as controllers for creating music by their scratching practice: the
audio signals really outgoing from the turntable by vinyl emulator are not anymore music on vinyl and DJs commands but just DJ commands. However, vinyl
emulator improves the potential level of outputs of this service characteristic by
extending the set of music sources (inputs) that DJs can control and let outgo
from turntables.

- Vinyl emulator improves the service characteristic "DJing" by augmenting,
  again, its inputs; the set of music sources and operations DJs can execute (albeit
  by using other technological devices) in their scratching performance. For
  instance, turntablists can use vinyl emulator in order to record and manipulate
  in real time their performance by software for special effects.
- DJs turntablists can minimize the externality "vinyl damage" due to its repetitive use as users can record in digital format, the music on vinyl and then play
  just the mp3 file in their future performances. These improvements in the outputs of some service characteristics of analog turntables augment the general
  effectiveness for DJ turntablists.

Users of vinyl emulator benefit of complementary digital technologies by which the outputs of some OTBP service characteristics and customer effectiveness are improved. Vinyl emulator makes turntablists' activity more efficient (e.g., by buying digital music at lower prices) and effective (e.g., by transporting music without effort). The vinyl emulator partially modifies also the way by which users experience and practice disk jockeying. However, the turntablists' static need of "feeling the vinyl" with their own hands is still satisfied, since they keep touching the format vinyl of the vinyl emulator.

A crucial condition for the success of vinyl emulator within the community of DJ turntablists has been that most of them clearly perceived how vinyl emulators provide the opportunity of continuing to use Technics SL-1200 without rejecting the advantages of digital sound and digital musical devices. The emergence of new technology shapes a new set of desired needs for DJs. These users resist to CDJs but are attracted by the benefits of digital music. The interoperability between the old turntables and digital music devices by vinyl emulator satisfies this evolution of customers' needs and becomes critical to improve the general performance of analog turntables and, as a consequence, the customer effectiveness of disk jockeying.

Other factors facilitating the diffusion of vinyl emulators between turntablists have been their promotion and direct use by many internationally recognized DJs (turntablists and not) and, furthermore, the possibilities of co-development provided by some firms to end-users. For instance, RANE manages a web-forum for product discussion in order to get feedbacks and suggestion for Scratch Live, its vinyl emulator ("Scratch Live Features Suggestions" is one of the main topics of RANE web-forum).

Since 2004, many firms developed and marketed vinyl emulators. Both large established firms (as Stanton and Numark) and smaller companies (as Torq) have commercialized CDJs and vinyl emulators. However, RANE corporation was the first to recognize the entrepreneurial opportunity of vinyl emulators even if, during the "analog turntables era," it was just a small company developing mixers and music equipment. Technics surprisingly did not develop its own vinyl emulator. The firm focused its attention on the larger DJs market by performing incremental innovations on its traditional analog turntables and launching its own CDJ (SL-DZ 1200). Therefore, this case shows, paradoxically, vinyl emulator has been a gateway device developed by some of the main competitors of Technics in order to protect and keep alive the Technics dominant turntable (SL-1200).

Vinyl emulator has not slowed the replacement of old analog turntables. This product is a niche artifact within the market segment of DJs and, therefore, it did not interrupt the process of technology change. Indeed, when vinyl emulators appeared on the market the process of substitution between CDJs and turntables was already at an advanced stage. However, vinyl emulators do support the continuities of preferences of DJ turntablists for old turntables. Furthermore, this vintage product enabled them to both (1) achieve the critical benefits of the digital sound paradigm and CDJs (music in mp3 format) and (2) support the adoption of the new technology for the implementation of their traditional core practices.

## 4.3   Film Scanner for Analog Photographers[3]

### 4.3.1   Analog Photography, Film Cameras, and Technological Change

In 1885, George Eastman created the modern photograph film technology that made analog cameras a convenient product for consumers. Also, in this case the core of technological change of this industry refers to the transition from analog cameras to digital cameras. Table 4.2 reports the chronology of the main events of this process of technological change.

Similar to turntable, film camera is just one of the components of a larger technological system that users adopt in order to perform the full process of analog photography. The technical sub-systems (each one made of various technical characteristics) of a typical analog camera are three: the film, camera body, and lens. The main typical service characteristics of traditional cameras are:

- main services: image capture, negatives reproducibility;
- complementary services: zooming and focusing, image manipulation;
- externalities: lens distortion and flare, film wear, and damage.

From the end-user perspective, the main technical difference between these two models of cameras is that analog cameras use as printing medium photographic films, sheets of plastic, to process (and manipulate to some extent by hands in the darkroom) afterwards in order to get pictures. Instead, digital cameras store images as digital files on memory cards, much more advanced (cheaper and more capacious in terms of number of pictures) than traditional photographic films. Afterwards, digital images can be printed and/or manipulated via other complementary digital products (e.g., software, PC). Digital cameras offer to customer innovative service characteristics (e.g., the "red eye effect" removal or the "anti-shake" function). Beyond these benefits for users in image capture, manipulation, and storage, digital images do not require complex and long processes of development or labs photofinishing in order to exploit one of the main service characteristic of analog cameras (film negatives reproducibility). Finally, digital cameras are usually cheaper and easier to use than film cameras.

Digital cameras started to be produced in large scale in Japan by Fuji (1988) and later in the United States (1991). Digital cameras surpassed analog cameras in sales in the United States in 2003 (see Fig. 4.2). Between 1991 and 2006, 83 companies entered into the American industry of digital cameras: 25 incumbent photography firms, 19 consumer electronics firms, 25 computer and computer peripheral firms, 9 de novo start-ups, and 5 diversifying entrants from unrelated industries (Benner and Tripsas 2012).

---

[3] This section extends some elements of the analysis developed by the author in Schiavone (2013b).

**Table 4.2** The chronology of the main events

| Years | Event |
|-------|-------|
| 1885 | George Eastman created the modern photograph film technology that made analog cameras a convenient product for consumers |
| 1988 | Fuji developed the DS-1P which is considered to be the first fully digital consumer camera to be sold |
| 1988 | Nikon launches its film scanner LS-3500, one of the first models in the product class |
| 2003 | Digital cameras surpassed analog cameras in sales in the United States |

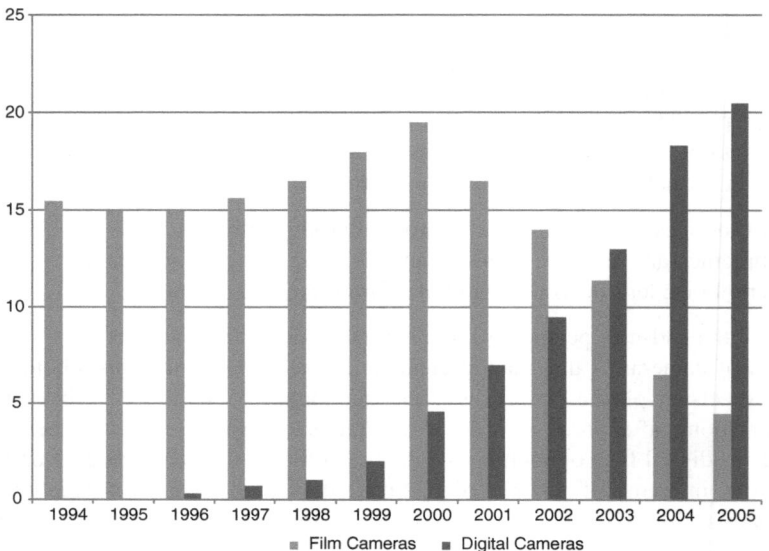

**Fig. 4.2** US Cameras Sales (adapted from Lucas Jr and Mein Goh 2009)

Such radical industry and technological changes implied that some dominant incumbent photography firms were not able to react properly to digitalization and almost failed, as Polaroid did (Tripsas and Gavetti 2000). Other companies (e.g., Kodak) had several problems in understanding this disruptive change, and as a consequence lost their dominant market positions Gavetti et al.(2005; Lucas Jr and Mein Goh 2009).

Throughout this period, the innovation activity of the industry's main firms was not solely focused on the new technological paradigm of digital imaging. For instance, the main established companies implemented the sailing ship effect approach by developing incremental innovations on analog film technology in order to support those users affording a part of (or all) these issues with digital cameras. Canon, Fuji, Kodak, Minolta, and Nikon developed and launched on the market the Advanced Photo Systems (APS) in the mid-1990s. APS was:

> an effort to boost analog film technology when the first consumer digital cameras were emerging in the early 1990s… The APS technology was basically a new film format of the existing silver ionic film technology. Its extra-thin film layer was entirely enclosed in a special compact cartridge allowing smaller camera dimensions… APS technology was first commercialized in 1996. However, its success was limited. With the rising popularity of digital cameras at in the end of the millennium, APS cameras hardly sold. Their commercialization was discontinued only a few years after their launch (Trauffler and Tschirky 2006, p. 6).

The aim of APS technology for these companies was to exploit the business opportunities coming from the resistance to technological change in some market segments of consumer photography, in which analog cameras and photographic films were still valuable and worthy of use for customers.

## 4.3.2  The Community of Analog Photographers

Despite the consumer photography industry quite quickly shifting from analog to digital photography, many professional and amateur photographers still continue to use their old film cameras. Analog photography is considered with "an indexical depth whereas digital photography as depthless" (Hand 2012).

Such depth is a critical pre-condition for the rise of a technology-based community of practice centered around analog photography. The core practices of the community are related to two basic dimensions:

• the knowledge and expertise in the use of film camera;
• the rituals and expertise about the photographic processing (negatives development and reproduction).

Before technological change "Camera clubs" were used to be important loci of social aggregation for the members of this community. Camera clubs are places in which analog local photographers show their works and share knowledge and information about their beloved practice and its artifacts. Competitions, workshops, and seminars between photographers are recurrently organized by these clubs. These social events reinforce the centrality of the OTBP and its practices for the community even nowadays despite new technologies (e.g., social networks or smartphones) providing more opportunities for virtual interactions between members. The widespread of internet supported these traditional places of social aggregation and interaction by the rise of specific online social networks for photography. Currently, there are several virtual communities celebrating and exchanging information, as occurs in traditional camera clubs, about analog photography and film cameras (e.g., Flickr, Instagram, APUG).

The survival of film cameras and communities of analog photographers over the digital era was supported also by various practical reasons inhibiting the transition of some groups of users to digital imaging (Runde et al. 2009):

- problems in the use (and with the form) of digital cameras and other digital devices;
- the prospective loss of pre-existing investments in skills and the cost of acquiring new ones (switching costs);
- demographics of different social groups, as shown by the "soccer mom" social group, in which the adoption of digital cameras was very slow.[4]

Another factor supporting the survival of analog photography was the belief of its community members that the outputs of film cameras have an aesthetic value that digital pictures cannot achieve. Overall, the debate about the competition between film cameras and digital cameras seems to be never-ending. Many skilled photographers claim that the quality of pictures taken with analog cameras is higher than that of digital pictures. Moreover, the latter are often not useful for all the practices of professional photographers (e.g., publication of their pictures in specialized magazines). The general opinion is that analog cameras and digital cameras complete each other as the former achieve a better quality results, whereas the latter are generally more flexible and convenient.

## 4.3.3  Film Scanners

After the rise of digital imaging, various companies developed and commercialized film (or negatives) scanner. This is a vintage product converting images on 35 or 120 mm films and negatives in digital files without the use of any intermediate printmaking. This product connects the output of old analog cameras (film negatives) with more devices of digital cameras (memory cards, computer hardware, and software). Film scanners are made out of two basic components:

- new technology-based device in which to insert the negatives to convert;
- special software to instal on PC.

Film scanners are connected via cable to PCs in order to transfer to them as digitalized images. These products also allow image manipulation, for instance the removal of scratches and film grain, as well as the improvement of color

---

[4] "A key aspect of the transition to digital imaging is that early adopters tended to be men, whereas the most active photographer in the US household is typically what the industry refers to as the 'soccer mom' in her family archivist role… The people who buy [the] majority of cameras in this country are women and the majority of people who take photographs and keep them, catalogue them and look after the memories for the family are basically women. The people that go out and buy techie, nice silver objects like this [points to digital camera] are men" (Runde et al. 2009, p. 17).

reproduction from film. Scanners benefit from the complementarity effects of film photography, consumer electronics, and personal computing (Benner and Tripsas 2012).

Film scanner improves the performance and customer effectiveness of traditional analog cameras in three ways:

- By increasing the outputs of the main service characteristic "film reproducibility": film scanner completely changes the steps of the analog photography process after taking pictures. Analog photographers reconvert their old film negatives and photos in digital format and store them on PC by scanners. The darkroom or photofinishing labs are not anymore the unique solutions to reproduce film negatives. And films are not anymore the unique type of support to store the image captured by analog camera. This vintage product provides an additional option to achieve these goals.

- By increasing the inputs and outputs of the complementary service characteristic "image manipulation": experienced analog photographers are used to manipulate images on films in the darkroom. Film scanners give users the possibility to exploit PC as new environment and device for image manipulation. The use of digital technology provides new inputs to this service characteristic, including many innovative effects, and tools much more sophisticated and powerful (e.g., removal of scratches), unachievable by the traditional process of film development.

- By minimizing the externalities "film wear and damage": the adoption of a workflow fully analog implies that photographers have to use always film negatives. However, films could be damaged and by external agents (e.g., warm, grain) and decrease the quality of the image captured over time. Film scanner allows the conversion of negatives in digital files whose original quality is "eternal" as they do not incur these risks.

This information about the improvement of performance and customers' effectiveness for the OTBP by film scanner is consistent also with the content of interviews, conducted by the author via e-mail, with a number of amateur analog photographers.[5] Most respondents argue that film scanner is useful and successful for their community. These products give users of film cameras the advantages of digital imaging (e.g., quick release of pictures, minor costs) without decreasing the quality of film images and experiencing the switching costs related to the substitution of their old cameras. However, despite various respondents appreciating the benefits of digital photography and scanners for film technology some experienced photographers

---

[5] The respondents (contacted for another ongoing research project of the author) are from seven nations (Australia, Canada, Germany, South-Africa, Sweden, USA, Vietnam) and are mainly also users of film scanner. The semi-structured questionnaire was designed for a larger research project. However, all relevant information about the dimensions under investigation in the present study was extracted, coded, and analyzed. All the respondents of this study are members of one of the most worldwide influential online forums about analog photography: APUG (www.apug.org), Analog Photography User Group.

criticize with harsh comments the use of film scanner by "newcomers." In this way, the conversion device decreased the overall quality of the skills and culture of analog community. For instance, an analog photographer interviewed argues that:

> With millions of them [film scanners] worldwide, suddenly there is a new breed of photographers who think they are 'experts' at what they do, while in fact a lot of their knowledge clashes a lot with what is known about film for traditional use. It is a culture clash of sorts, of majestic proportions.

Therefore, on the one hand, film scanner supports the survival of the community to technological change by improving the performance of analog cameras and allowing the rise of this specific niche of OTBP users. On the other hand, this culture clash can be considered as a sort of unexpected negative effect of the vintage product on the community's original identity.

The need for converting film negatives is common across various segments of the analog cameras market and it is not felt only by those social groups slowly adopting new technology (e.g., soccer-moms). Film scanners became valuable artifacts for three segments of the consumer photography market. The first two segments existed already before technological change while the third emerged over the digital era.

First, this vintage product is very useful for professional photographers believing in the superior quality of analog cameras. Professional film scanners quickly became an alternative way to darkroom for many photographers in order to perform some of their practices and activities.

Second, film scanners are used also by amateurs photographers interested in the conversion of their old film negatives and photos into digital format. For instance, in the community web-forum Thephotoforum.com," an old film camera user explicitly states that and asks for suggestions about affordable scanners to buy:

> I have, like many photographers have, a box of old photo's and negatives, What I would like to do scan the negatives, can anyone recommend a inexpensive Photo Negative?[6]

Third, film scanner is critical to attract, even after technological change, new users to the community of practice of analog photography. For instance, lomographers widely use conversion devices to publish online film negatives taken by analog cameras. Lomographers are "amateur photographers [that] take spontaneous snapshots in everyday situations. They use obsolete Russian cameras (lomocameras), and would for example "shoot from the hip," instead of looking through the viewfinder when taking pictures" (Ljungblad 2008, p. 42). At the beginning of 2013, the first film scanner for lomographers started to be developed. Its key characteristic is that the film scanner is not connected, as usual, to PCs but to smartphones (see Fig. 4.3).

Film scanners are technological artifacts mainly based on new digital technology. It is for such a reason that this converter was also developed and commercialized by new entrants in the consumer photography market, after technological

---

[6]  http://www.thephotoforum.com/forum/photography-equipment-products/197281-scan-old-negatives.html (accessed 8 November 2011).

**Fig. 4.3** Film scanner
for lomographers (source:
Lomography)

change (e.g., electronics and computer companies as Epson or HP) developed and commercialized this product. Nikon was one of the first established companies to commercialize film scanners by launching the model LS-3500 in 1988. Also, Minolta and Canon put great effort in the development of this innovation. Fuji, Kodak, and Polaroid started developing film scanners afterwards.

Despite the success and effectiveness of this vintage product for customers, analog photographers interviewed outline two critical complementary evidences about the scarce development of film scanner by companies. First, the rapid pace of technological change did not push companies in developing further this product of interconnection. In this light, for instance one respondent argues that:

> with the explosion of electronic forms of photography, scanning machine manufacturers seemed to stop research and development into better scanning instruments and software. This is usually felt more with people using sheet film as their film size and possible information that can be extracted, often exceeds the scanning devices' ability.

Second, various respondents agree that film scanner is nothing more than a niche product and, thus, it is not a relevant source of profits for companies. For instance, another photographer reports that:

> Personally, scanners make my use of film viable… [however] I suspect that scanners are not a reliable money earner for manufacturers, noting the discontinuance of all the mid-range (e.g., Nikon 9000-class) scanners.

## 4.4   Conclusions and Implications

These two case studies provide a number of valuable evidences about vintage innovation as corporate approach.

Vintage products are basically new technology-based converters having a high degree of technical complexity. The integration of different technologies within vintage products is necessary in order to achieve modularity and interoperability between different technological systems. The technical structure of vintage products developed by companies tends to be more complex and diversified than vintage products developed by users (see Chap. 5). Indeed, companies have more resources and capabilities to invest in R & D and manufacturing than standard end-users of OTBPs. However, no particular differences should emerge between vintage innovation by companies and end-users in case this phenomenon is based on emulation via digital technologies.

Vintage products improve the performances of OTBPs and customer effectiveness by networking the OTBP and new technology and keeping the same technical characteristics of the OTBP. The lack of technical changes in the OTBP is the base for the preservation of the users' traditional key technological practices. This condition decreases the switching costs of old technology users, critical barriers to the adoption of new technology. Therefore, vintage products provide very important advantages for companies serving old technology-based communities but also commercializing new technology-based products. However, companies could achieve this benefit just if vintage products fulfill some requirements. The cases show users might tolerate changes in some practices (e.g., hybrid workflow by PC for analog photographers) and might not appreciate changes of others (e.g., scratching by CDJs for DJ turntablists). Moreover, vintage products affect all the types of services provided by old products without distinctions. Companies have to understand properly which are the real links between changes in service characteristics of OTBPs and the users' practices related to old technology. In order to improve effectively customer effectiveness, firms must develop vintage products affecting only OTBP service characteristics that are relevant for users and that do not reshuffle, by improving, the key practices and knowledge of their community.

The cases outline two interesting evidences for companies in the domain of technology management. First, film scanner and vinyl emulator have similar technological characteristics and goals. Companies perform vintage innovation by combining both traditional physical gateway devices and software for emulation. These evidences suggest companies can develop effective vintage products just by exploiting digital technologies. They are a critical element for the innovative type of extension of the OTBP service characteristics provided by vintage products. Moreover, companies can exploit ICT to provide a virtual architecture to the communities of vintage users for co-creation of these products. Second, vintage products emerge when the OTBPs are experiencing their phases of decline. This condition implies that new competing technology is already dominant and old product users are aware of its benefits for their traditional technological practices.

Referring to innovation strategy, two interesting evidences emerge from the case studies. First, the implementation of vintage innovation is not exclusive of some specific type of organization. Both incumbent companies and new industry entrants, small and large firms can implement this approach. Furthermore, vintage products are developed also by companies without any specific experience in OTBP manufacturing and marketing. Second, vintage products are often a single component of a larger innovation strategy in which both old and new technology are developed and commercialized. In terms of R & D, the implementation of the phenomenon of vintage innovation by firms leads to an innovation approach that can be named "technology reverse" (Schiavone 2013a) as companies basically do not look forward and do R & D to increase technical progress but backwards to establish compatibility between technologies of different generations.

The marketing strategy of vintage products is likely to be quite standard. Internet and digital technologies are critical means for commercialisation of vintage products. Internet websites can serve as direct channels of selling and distribution of vintage products (at least for their software components). At product level, vintage products imply the integration of both intangible services and tangible components. Each of these components might be a strategic source of personalization for customers.

The social phenomenon of lomography shows vintage products have a great potential to discover, be purchased by, and provide effectiveness to new niches of OTBP users. The maturity, fascination, and prestige of the old product and its technology support also the research by companies for new "second generation" customers. The value proposition of vintage products for new generation users of OTBP should refer to the possibility to become, by using these products, skilled fascinating, and retro-chic as traditional customers of OTBPs.

Drawing on these evident fascinations of OTBPs for people belonging or aspiring to become members of these communities of practice, marketing communication of vintage products should highlight the depth and value of the community practices based on old products. The brand awareness related to declining dominant design should be another resource that incumbent companies might exploit for marketing vintage products.

# References

Benner MJ, Tripsas M (2012) The influence of prior industry affiliation on framing in nascent industries: the evolution of digital cameras. Strateg Manag J 33:277–302

Brewster B, Broughton F (2000) Last night a DJ saved my life. The history of disk jockey. Groove Atlantic Publishers, New York

Dosi G (1982) Technological paradigms and technological trajectories. Res Policy 11(3):147–162

Eisenhardt KM (1989). Building theories from case study research. Acad Manag Rev 14(4):532–550

Faulkner P, Runde J (2009) On the identity of technological objects and user innovations in function. Acad Manag Rev 34(3):442–462

Flyvbjerg B (2011) Case Study. In: Denzin NK, Lincoln YS (eds) The sage handbook of qualitative research, 4th edn. CA, Thousand Oaks, pp 301–316

Gavetti G, Henderson RM Giorgi S (2005) Kodak and the digital revolution. Harvard Business School Case, Boston

Hague P, Hague N, Morgan CA (2004) Market research in practice. A guide to the basics, London

Hand M (2012) Ubiquitous photography. Polity, Cambridge

Lippit TM (2006) Turntable music in the digital era: designing alternative tools for new turntable expression. NIME Conference, Paris

Ljungblad S (2008) Beyond users. Doctoral Thesis, Department of Computer Science, Stockholm University, Grounding Technology in Experience

Lucas HC Jr, Mein Goh J (2009) Disruptive technology: how Kodak missed the digital photography revolution. J Strateg Inf Syst 18:46–55

Runde J, Jones M, Munir K, Nikolychuk L (2009) On technological objects and the adoption of technological product innovations: rules, routines and the transition from analog photography to digital imaging. Camb J Econ 33:1–24

Scapens RW (2004) Doing case study research. In: Humphrey C, Lee B (a cura di). The real life guide to accounting research, Elsevier, New York, pp. 257–279

Schiavone F (2013a) Vintage innovation: how to improve the service characteristics and customer effectiveness of products becoming obsolete. IEEE Trans Eng Manag 60(2)

Schiavone F (2013b) Innovation approaches for old products revitalisation after technological change: the rise of technology reverse. Forthcoming Int J Innov Manage

Smith S (2000) Compositional strategies of the hip-hop turntablist. Organised Sound 5(2):75–79

Smith S (2007) The process of collective creation in the composition of UK hip-hop turntable team routines. Organised Sound 12(1):79–87

Souvignier T (2003) The world of DJs and the turntable culture. Hal Leonard Corporation, Milwaukee

Trauffler G, Tschirky H (2006) Sustained innovation management. Palgrave Macmillan, New York

Tripsas M, Gavetti G (2000) Capabilities and cognition, and inertia: evidence from digital imaging. Strateg Manag J 21:1147–1162

Yin R (1994) Case study research: design and methods. Sage, CA

van den Dobbelsteen M (2008) The tables are turned, MSc Thesis, University of Amsterdam, Amsterdam

Weber M (1949) Objectivity in social science and social policy. In: Shils EA, Finch HA (eds) The methodology of the social sciences. Free Press, New York pp. 50–112

# Chapter 5
# Vintage Innovation by Users

**Abstract** This chapter reports two case studies on vintage innovation developed directly by communities of users of old technology-based products. This type of vintage innovation is based on the so-called phenomenon of user innovation, the development of an innovation by intermediate users, or consumer users. The cases reported in this chapter are illustrative and based on primary data, collected by e-mail interviews to old technology users, and secondary data collected by scientific references, and online resources. The unit of analysis is the improvement of OTBP performance and customer effectiveness via vintage products after the emergence of new competing technologies. Three dimensions are described for each case study: the OTBP and the process of technological change, the community of practice, and the vintage product. The two case studies are about vintage products for ham radios and arcade videogames, developed by global communities of aficionados users, nowadays communicating and interacting mainly by the Internet. The chapter ends by summarizing the main conclusions and managerial implications of this specific form of vintage innovation.

**Keywords** User innovation • Radio-amateurs • VOIP • Arcade videogame players • MAME 32 • Emulation

## 5.1 Introduction

This chapter reports two case studies on vintage innovation developed directly by communities of OTBPs' customers. This type of vintage innovation is based on the so-called phenomenon of user innovation, the development of an innovation by intermediate users or consumer users. Users can undertake this process for

F. Schiavone, *Communities of Practice and Vintage Innovation*,
SpringerBriefs in Business, DOI: 10.1007/978-3-319-01902-4_5,
© The Author(s) 2014

innovating both industrial products and consumer products. Two technical trends in many industries support user innovation (Von Hippel 2003):

• the improving design capabilities (innovation toolkits) in ICT industries that make easier and more accessible this form of innovation;
• the improving ability of individual users to combine and coordinate their innovation-related efforts by the Internet.

A typical motivation behind this phenomenon are the incentives and expected benefits that user innovators receive by their innovation activities (Bogers et al. 2010). Companies sometimes do not perceive users' needs (or do not consider them economically worthy of interest) and, thus, do not develop any specific product or solution. Therefore, the only way for users to satisfy their needs is to innovate by themselves. Another explanation of user innovation lies in the innovation-related costs which are lower for users due to information stickiness and their deep expertise in the usage and technical characteristics of the product.

The great efforts and complexity behind user innovation make this process very hard to develop for just one person. Typical examples of communities of technological users performing user innovation are the developers of open source software and windsurfers (Von Hippel 2001). As argued by Von Hippel in several of his seminal articles, this phenomenon is widespread with the rise of digital technologies and the Internet. This explains why user innovation is very often not developed by single individuals but by large virtual (online) groups of users that interact in the various phases of the innovation process over time.

The cases described in the following show how, by digital technologies, groups of technology-skilled users can be, under given conditions, more than competitors of industry firms. These users can even be sources of business and value for incumbent companies as long they realize the economic value that can be gained by supporting user innovation. Also, in vintage innovation by users there is the close connection between the community key practice and the OTBP in decline. The uniqueness of these community needs, apparently without any commercial value or appeal for companies, pushed users to develop by themselves vintage products improving the performance of their beloved OTBPs.

As well as the two case studies in Chap. 4, the cases reported in this chapter are illustrative and based on primary data (collected by e-mail interviews to old technology' users) and secondary data collected by scientific references and online resources (Hague et al. 2004; Scapens 2004; Yin 1994). The unit of analysis also for the cases reported in this chapter is the improvement of OTBP performance and customer effectiveness via vintage products after the emergence of new competing technologies. Three dimensions are described for each case study: the OTBP and the process of technological change, the community of practice, and the vintage product.

The following reports two case studies of two OTBPs (ham radio and arcade videogames) innovated by global communities of aficionados users, nowadays communicating and interacting mainly by the Internet. The chapter ends by summarizing the main conclusions and managerial implications of this specific form of vintage innovation.

## 5.2 Multimedia Software for Radio-Amateurs[1]

### 5.2.1 History and Technological Change of Ham Radios

Ham radios (or amateur radios) could be considered as the core technological product of the first social network in the history. Ham radios have been performing for more than a century functions similar to Internet websites such as Facebook, Twitter, and LinkedIn today: they allow communication and establish social ties between different groups of persons, very often unknown to each other, often from different and distant geographical areas. The first official reference about amateur radio is by Hugo Gernsback, Luxembourgian-American inventor and writer, in 1909. In his book "The First Annual Official Blue Book of Wireless Association of America,"[2] Gernsback reports the list of radio stations held by the American Army and private citizens.

Ham radios can be of various types: base (or fixed) radios, handheld radios, mobile radios, and portable radios. Amateur radios allow users two modes of transmission. The first modality is Frequency Modulation (FM), which offers high quality communication. The second modality is Single Sideband (SSB).

A large amount of official publications and reports self-produced by radio-amateurs describe the main technical characteristics, service characteristics, and the overall technological system of typical ham radios. Technical characteristics might change across the different types of radios. However, some common technical elements of every ham-radios are antenna, microphone, display, memory for channels storing, system for signal processing, system for power supply. Ham radios provide to users various service characteristics:

- main services: communication with other radios;
- complementary services: signals receiving, signals outgoing, scanning, programming and configuration, dual receive;
- externalities: vulnerability to external agents.

Referring to externalities, bad weather conditions (e.g., storms) and long distance between radio operators affect negatively the quality and success of communication between old ham radios. Moreover, despite this OTBP is still nowadays very fascinating, communication via ham radios is often unreliable also due to the incorrect configurations and manipulations of equipment by amateurs (Bogdan and Bowers 2007).

Over the past two decades, the rise of in more sophisticated technological systems of real time wireless communication (e.g., telephone, Internet) undermined

---

[1] This section is based on and extends some elements of the analyses developed by the author in Schiavone (2011) and Schiavone (2012).

[2] A digital copy of the original book by Gernsback can be downloaded at the following weblink: http://www.seas.upenn.edu/~uparc/documents/First%20Annual%20Official%20Wireless%20 Blue%20Book%20-%201909.pdf (Url retrieved: 21-06-2013).

the use and utility of ham radio in many of its historical fields of application. As argued above, communication via amateur radio entails many technical restrictions and problems that new technologies easily solve. The higher quality of communication through new technologies decreased the social relevance and use of ham radios. Moreover, technical evolution in the telecommunications sector generated a strong competition between different systems of communication for the acquisition of electromagnetic spectrum frequencies. Companies would in fact reduce the size of spectrum (the frequencies) allocated to old ham radio in order to improve the size for more advanced systems (e.g., wireless communications). This industry change makes ham radio use more problematic and less performing.

## 5.2.2  The Community of Radio-Amateurs

The case of radio-amateurs is probably one of the best examples in which the OTPB, its practices, and structures (à la Orlikoskwi 2000) cannot be detached by its community of users. Radio-amateurs are probably one of the oldest technology-based CoPs in the recent history. The size and boundaries of this community are evidently depending on the pace of technological developments of science and, in particular, radio technology.

In the early 1900s, the first communities of radio-amateurs were local or regional. Nowadays, complementary technologies (as informatics) support the expansion of communities that, at least in theory, can reach global dimensions. The practice of this community was widespread for many decades until new and better performing technologies for long-distance communication emerged. Before technological change, radio-amateurs played an important social function in the past. For instance, ham radios were one of the key informal channels of news communication and dissemination during the first and second world wars. Still nowadays radio-amateurs play an important role when war emergencies or natural disasters occur, as for terrorist attacks in New York in September 2001 or the Asian Tsunami in December 2004. The amateurish nature of this community implies that its members do not receive any compensation for their activities. Today, more than 6 million radio-amateurs are estimated to work worldwide.

The official access to the community is regulated in many countries through a public examination after which the successful candidates get an official national license to transmit via ham radios and a personal radio amateur code ("call sign").

Today, the practice of radio-amateur activity is mostly just a hobby and seldom keeps a social function (such as the extra-ordinary situations cited above). Apart from the technical limitations cited above that are implicit to the OTBP, however, technological change per se never created a real risk of technological substitution within the community or a risk of disappearance of the whole social group.

The book by Kristen Haring (2007) is a valuable source of information to understand the reasons for such resistance to technological substitution and the relevance of culture and practices, exclusively based on radio technology, for the members of this community. Technical culture provides to the community of radio-amateurs a technological identity by which they feel as members of a specific and special socio-technological group. In this light, Haring (page XV 2007) argues that:

> Hams deliberately set themselves apart by developing a community and culture tied to radio technology. They articulated technical values, goals, and practices different from those of non-hams and used adherence to this way of thinking to judge group members. That is, radio hobbyists formed their own 'technical culture', a culture built around and establishing an ideology about technology.

A very interesting example of the richness of practices and communal intangible resources within this community relates to the wide use of the Morse code between users to communicate secret information and create "privacy in public." For instance, radio-amateurs were often used to communicate secretly across a crowded room with a fellow spelling his call sign by the Morse. Therefore, the learning of proper (and improper) sending techniques of messages with Morse code became a relevant bundle of knowledge for some users.

Speaking habits are rigorously regulated within this community. Radio-amateurs have various ethic codes and rules which should be respected by the members of this community. The basic principles of this community are (The International Amateur Radio Union 2010) the social feeling, brotherly spirit and brotherhood, tolerance, politeness, and comprehension.

These are just a few of the many examples of the existence of the learning efforts and dynamics, social norms, formalized codes, and common knowledge on which this community is built. All these intangible resources of the community supported the survival and use of ham radios even in the digital era. Other motivations of resistance to the substitution of this OTBP within the community of amateur radio operators are its important historical tradition within society, its institutional recognition, and the strong set of social ties, formalized and informal, on which the community is based.

However, many radio-amateurs recognized the potential benefits of new wireless communication technologies for their practice as soon these technologies emerged. The role of change agents was critical for orienting the community toward an ambivalent response to change, merging resistance, and openness to new technology, by radio-amateurs (Schiavone 2012). These members were not necessarily opinion leaders of the community before technological change. The main capability of change agents regards the satisfaction of the latent needs of the community (e.g., to fix technical limitations of communication by traditional radios) by exploiting and supporting new technology.

### 5.2.3 Multimedia Software

Over the past 15 years some radio-amateurs developed multimedia software and hardware linking together traditional ham radios to Internet broadband and the VOIP (Voice over Internet Protocol) system in order to overcome the traditional technical limitations of radios and improve the scope of messages sending and receiving (Fig. 5.1). Many specialized websites for radio-amateurs describe in detail how such multimedia technological systems establish a collaborative relationship between these two competing technologies (radiophony and Internet):

> Ham radio has always relied on FM repeaters, large radio towers that act as base stations for accessing the radio network from home. By attaching an Internet-connected PC to these repeater stations, people can communicate with the repeater using VoIP... Users can connect their ham radio transceivers to their PC sound card and use the computer software to search for available repeater stations across the world (Valdes and Roos 2001).

Thus, the scope of radio-amateurs practice is no longer limited just to the repeaters nearer to their place of transmission. Issues of distance between operators and weather conditions are so fixed by new technology. These repeaters are not of public property but radio operators own, use, and maintain them.

The first multimedia system connecting traditional radios with VoIP was IRLP (Internet Radio Linking Project, http://www.irlp.net). A Canadian radio-amateur (David Cameron, VE7LTD) started the project in 1997. He built and tested a combined hardware and software (running on Linux) connecting two extremely distant radios over the Internet. IRLP is currently used all around the world.

Echolink (http://www.echolink.org) serves the same end of IRLP. This project was started by (Jonathan Taylor, K1RFD), an American radio-operator, in 2002. Nowadays, the development of Echolink is not based anymore on the commitment of a single individual (as IRLP) but on the collective and co-ordinated activities of several members of the community. More than 200.000 amateur radio operators in 151 countries worldwide currently use this software. The average number of users

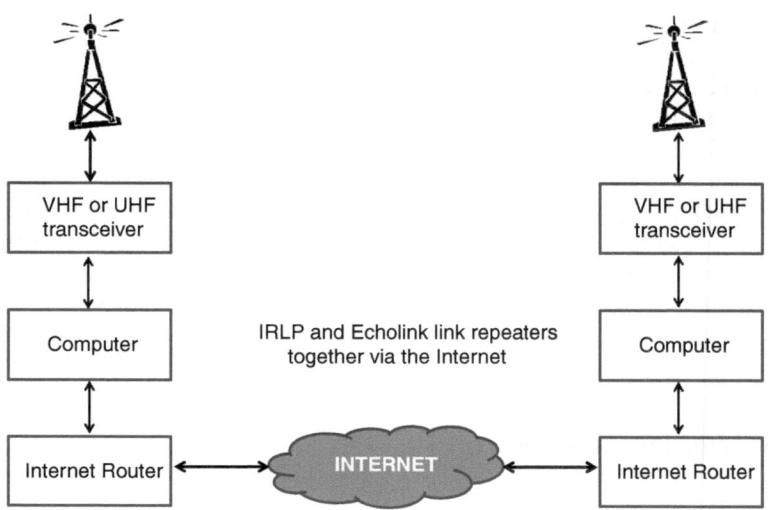

**Fig. 5.1** The Echolink and IRLP systems

connected to the software at any given moment is 5,200. Echolink can establish two different types of connections between radio-amateurs:

- PC to PC: The Radio Amateurs Use the Echolink Software Installed on the PC to Communicate to Each Other.
- Sys op mode: The radio operator communicates between PC and transreceiver. There are two basic modes. The first is the "simplex link" mode in which an Echolink user connected to a node via Internet connection and a ham operator is connected to the same node though the air. The second is the "repeater link" mode, in which the node is connected to a repeater.

Frequently, radio-amateurs using VOIP are registered users on both IRLP and Echolink. Both are open source software, free to use and download from the web. However, the project leaders accept spontaneous donations from users adopting their software. Some radio-amateurs gain value from these systems technology by drafting and trading software manuals and guides. The programmers of both these softwares developed applications for smartphones over the past few years.

These innovative multimedia systems received great attention in the community debate. Some purists inside the community believe that use of such systems denaturalizes the radio-amateurs practice. For instance, Echolink can establish communication between just two PCs and skip completely the use of traditional ham radios. Therefore, this alternative use of the vintage product might improve the customer effectives by improving the performance of the practice rather than the OTBP.

However, the use of the old technological product remains unchanged even by using IRLP or Echolink (unless, as argued, the radio operator decides to connect the system not directly to his traditional radio but to PC). This type of software improves at least three service characteristics of ham radios:

- The outputs of the main service "communication with other ham radios": the interoperability with VOIP technology improves the amount of communication options between two radio-amateurs (e.g., the sys op mode). Moreover, Echolink and similar software improve even the quality of the signal (thanks to VOIP technology).
- The outputs of the complementary service "scanning": Multimedia software improves the power of ham radios in scanning and searching for repeater stations not reachable by the standard technological configuration.
- The outputs of the externality "vulnerability to external agents": multimedia software overcome the usual barriers of ham radios and make communication between radio-amateurs less affected by external conditions (as the weather) or geographical distance.

As a consequence, the final level of effectiveness for radio amateurs still using traditional ham radios improves greatly.

The analysis of the improvement of performance by vintage products cannot be developed separately from the understanding of the process of user innovation.

To this end, the author performed a number of personal interviews with a small sample of Italian radio-amateurs and Jonathan Taylor (IRLP software developer) over time.[3] Questionnaires responded by a number of community change agents (namely, those radio-operators who supported the diffusion of new technology into the community) provide relevant information about user innovation and the set of responses of this community to industry technological change. In general, all the key informants agree on the relevant benefits of multimedia software for the performance of the traditional radio-amateurs' practice.

Four key informants stress their responses that orientation to experimentation, a critical feature of the radio-amateurs community, played an important role in the development of multimedia software and adoption of new technology. Communication is a critical way by which these technological users do experimentation (Bogdan and Bowers 2007). The willingness of users to learn and apply new knowledge is the background condition for the activities of development and promotion of novelties into the community by some users working as "change agents" within the community. Internet is complementary to informatics, which greatly rose over the past years. The strong technological complementarity between electronics and informatics was a crucial condition to create a positive technological context for the development (actually, by few ones) and the adoption (much larger) of Internet and multimedia software by radio-amateurs communities. Key informant 5 argues that these two "pastimes" (ham radios and computers) have common technical roots and share an interest in experimentation. This condition made quite easy for some skilled radio-amateurs the exploitation of Internet for improving the conventional practices of radio-amateurs. Key informant 4, in his reply to the question "Did technological change created problems of maintenance for old ham radios?" notes the paradox that:

> ... nowadays the use of radios is increasingly determined by software and, thus, it implies to radio-amateurs the learning of an in depth knowledge of informatics and specialization in signal processing.

In relation to the process of user innovation, Jonathan Taylor replies to the question "Who is the typical developer of such kind of software? Does he hold more specialized and critical technological knowledge than a classical radio-amateur?" by stating that:

> Relatively few hams have become involved in actual development of software. Those that have tend to be people who have previous (or current) experience in developing general-purpose software (as opposed to ham radio software), either as a hobby or as profession, and are applying the same skills to amateur radio software development. For the most part, this is a different set of knowledge and skills than those who pursue pure amateur radio.

Despite the process of innovation development being concentrated in the hands of a few radio-amateurs, many community members (in Italy and across the world) greatly supported the adoption of VoIP. For instance, key informant

---

[3] These interviews are the empirical base of two articles published by the author over the last few years (Schiavone 2011, 2012).

2 independently developed a website and a users' guide for multimedia software in order to stimulate the social contagion of this vintage product in the Italian radio-amateur community. Overall, the diffusion of multimedia software as IRLP and Echolink were boosted also by the typical desire of experimentation of many users and the tangible benefits on the performance of the OTBP.

The use of VoIP reshuffled the social structure of the community and generated new subnetworks. In this light, key informant 5 notes that the adoption of new technology created new small niches of users within the community. However, Key informants 1 and 2 note that some (but definitely not all the) Italian associations of radio-amateurs were skeptic about the use of VoIP and Internet for communication by ham radios. The final outcome, according to key informant 2, was that in this European country the adoption of multimedia software was lower than other countries.

In sum, user innovation and ambivalent responses to technological change within the radio-amateurs community are likely to rely on two conditions: (1) the presence within the community of some skilled members able to perceive the benefits of new technology for old technology and (foremost) to exploit them by developing vintage products; (2) the general orientation of the community for experimentation, learning, and exploitation of new technology's knowledge. Some incumbent companies exploited these conditions and the innovation-related efforts of this community. For instance, Yaesu developed interfaces, software and hardware specifically designed to support Echolink and IRLP.

## 5.3   Emulators for Arcade Videogames Players[4]

### 5.3.1   Arcade Videogames

Informatics is a technological field rich with possibilities to develop vintage innovations. A good example comes from emulation software of old arcade videogames (e.g., Pac-Man, Tetris, and Space Invaders) on modern home PCs.

An arcade videogame is a coin-operated entertainment machine usually installed in public businesses, such as restaurants, bars, and particularly amusement arcades. In 1961, MIT student Steve Russell created Spacewar, the first interactive computer game. In 1968, the first videogame was patented. The inventor was Ralph Baer, a German-American engineer who after almost 20 years of work (his first attempt was in 1951) created an interactive game that can be played on a television screen. Videogame industry, after the "great crash" in 1983, had a great expansion over the past three decades. The 1980s were the first years of the modern videogame era with the release of many popular and successful games which were played by users in both public game centers and, after the widespread of home computers, their own houses. This period is commonly considered as the

---

[4] This section is based on and extends some elements of the analysis developed by the author and his co-worker in Schiavone and Agrifoglio (2012).

**Table 5.1** The chronology of the main events

| Phase | Period | Description |
| --- | --- | --- |
| Before the games | 1889–1970 | Antecedents of videogames industry. For instance, in 1889 the future Nintendo was founded. In 1970 the Arcade-game manufacturer Nutting Associates starts to develop a console and Computer Space, the first arcade videogame self-produced by Nolan Bushnell. |
| The games begin | 1971–1977 | The first machines and consoles are developed and sold. Bushnell founds Atari and develops the tennis videogame "Pong." |
| The golden age | 1978–1981 | The most famous arcade videogames of the history arise: Football, Space Invaders, Pac-man. Home consoles start to be sold. |
| The great crash | 1982–1984 | The high intensity of competition and sudden decrease of sales make failing many videogames companies. The "Pac-Mania" rises. Commodore launches Commodore64. |
| Videogames are back | 1985–1988 | Alex Pajitnov designs Tetris. Nintendo launches NES and Super Mario Bros. |
| The home market expands | 1989–1992 | Nintendo launches the Game Boy console. NES and SEGA consoles expand to 16-bit. Street Fighter II is launched on the market. |
| The 32-Bit era begins | 1993–1997 | Systems become more powerful (32-bit). The competition between SEGA and Nintendo increases. Sony enters into the market with Playstation, one of the most successful consoled in the history. |
| The modern age | 1998–1999 | Also Microsoft enters into the industry with its own console, X-Box. |

"golden age of arcade videogames." Drawing on the description of Herman et al. (2002), Table 5.1 reports the key phases of the history of arcade videogames.

This reconstruction of the history of videogame industry shows the complexity to identify a clear and univocal milestone or trajectory of technological change. The firms' continuous orientation to innovation and development of new videogame genres and gaming modalities outlines a number of continuous "microchanges." Nowadays, arcade videogames and videogames centers still exist but what mainly changed over time is the technical nature of videogames. A critical advancement in videogame industry has been the transition from 2D and 3D videogames to "virtual reality" games, running only in PCs equipped with more sophisticated and powerful hardware (e.g., video cards, sound card, and RAM memories). In other words, technological change in the videogame industry is related to the shift (in both public game centers and players' private homes) from old "bit-based" platforms to the new generation of more powerful computers. Nowadays, virtual reality is the leading design in the videogame industry. For at least 10 years, virtual reality videogames have been replacing quite easily old videogames at both game centers and home players. The diffusion of home consoles as Playstation or X-Box has been extensive since the 1990s. Scholars recognized at least six different generations of home videogame consoles from 1985 to 2001 (Corts and Lederman 2009). The growth of home gaming has been another important trajectory of change for the industry which contrasted the traditional model of fruition and the nature itself of arcade videogames.

Videogames are particular products as they are ontologies, products without physical form (Faulkner and Runde 2009). Therefore, their technical characteristics are detached by any tangible component of the product. Typical technical characteristics of videogames are programming codes and files size (in terms of bytes). The performance of videogames can be decomposed in terms of mechanics, dynamics, and aesthetics (Fernández-Vara 2009). In the present analysis, the service characteristics of traditional arcade videogames are classified as follows:

- main services: gaming, multiplaying;
- complementary services: human-software interaction, interoperability with external devices (cabinets);
- externalities: risk of game crash.

Users can play arcade videogames just in public games centers by the interconnection of the software with external PC-centered technological systems. This form of gaming is based on the interaction between the player and the videogame (the player sends a command to the software which processes it and reacts). Videogamers might extend usual technological systems of PC with specific devices (e.g., joysticks) in order to improve videogames performance and usability just when they play at home. Therefore, the degree of personalization of technological systems and videogame-player interactions is null for arcade videogames played in public game centers.

## 5.3.2  Communities of Nostalgic Videogamers

After the rise of more sophisticated and complex videogames (e.g., 64-bit systems and virtual reality games) and forms of gaming (e.g., online gaming and home consoles) a number of communities of aficionados of old arcade videogames emerged quite soon.

Various motivations explain the reasons for which so many players across the world felt the nostalgia of old arcade videogames and created, for instance, virtual communities celebrating them. One key aspect depends on the inner nature of arcade videogames. Many arcade videogames provided to players the incentive of neverending games. Players were used to play as long their character was not "killed" by enemies. "Skillful players were rewarded with longer games" (Murphy 2013, p. 45). This characteristic of arcade videogames made many players very expert and reluctant to lose their skills and competences after the decline of their beloved videogames. Indeed, playing and winning in these videogames was self-satisfactory for many users despite these skills obviously not useful in their real life. Traditional cabinets usually allowed to play simultaneously more than two or three (very rarely four) users. Videogamers were used to engage real videogame battles in which they were opponents or partners depending on the rules of the played videogame.

Another element of strength of arcade videogames is the special experience that many players used to feel during games. Arcade videogames were "consumed" by interfacing with fascinating cabinets (on which the videogame, joystick, and other control devices were installed) in public game centers. This experience was for any user much richer and memorable than playing at home on his own computer or console.

Another feature of arcade videogames making them special to users is the possibility offered to players to save high scores. This function (not allowed by MAME and other emulators) created a sort of competition between the players interested to gain honor and notoriety within their usual game center (Amis 1982). In other words, the community of arcade videogames players was (and actually still nowadays by emulators is) segmented in many subcommunities, each of them focused around a specific famous videogame or videogame gender (e.g., shoot 'em up, sport games, role playing games).

Finally, arcade videogames have always provided social pleasure and nice moments of social aggregation to gamers. In the past social aggregation was achieved, for instance, by playing simultaneously with two or three friends to the same videogame in the same game center. This characteristic is closely connected with the centrality of the practices, skills, and knowledge that users need to play. People interact, exchange information and tips, observe others' playing techniques, and engage in local and/or virtual communities about arcade videogames in order to master successful video-gaming practice. For instance, communal intangible resources of these communities were the techniques to score goals more easily when playing a soccer arcade videogame as Exciting Soccer or Mexico 86. This type of learning might be very relevant for videogamers. Indeed, the acquisition and improvement of gaming practices sometimes are not important just for communities of gamers playing for ludic purposes. These techniques can have relevant implications for people in some social groups playing videogames to practice virtually on how to react and/or behave in real-life. For instance, US mariners were used to play videogames as "Doom" or flight simulators to feel and practice themselves in complex war situations (Poole 2004).

A key feature of these communities is the high extent of interaction between members. For instance, the manifesto of MAMEWorld community (focused on the vintage product described in the following subsection) clearly reports that their website is:[5]

> … not a cemetery of game news. We have a number of busy forums, chat, and we encourage the contributions and co-operation of our visitors with regard to all aspects of MAMEWorld. Check out the Fanstuff Page, the Rips Page, the Tunes Page, the reviews Pages, MameTesters, etc., and all our hosted sites. Many visitors contribute regularly and share their enthusiasm for MAME, the games and MAMEWorld. It is one of the most rewarding experiences about running this site to receive so many enthusiastic

---

[5] http://www.mameworld.info/net/mission.html (url retrieved: 18-06-2013).

contributions and to see the pleasures which MAME creates for so many people. And we respond personally to all requests, proposals or questions we get.

The MAME community's aim is to preserve gaming history by preventing vintage games from being lost or forgotten. About this, the official website of MAME project (http://mamedev.org/) reports that:

> MAME is done both for educational purposes and for preservation purposes, in order to prevent many historical games from disappearing forever once the hardware they run on stops working.

### 5.3.3  MAME and Other Emulation Software

Vintage products contributed to the enlargement and reinforcement of many communities of arcade videogames players. To date, MAME is probably the most famous emulator software running old arcade videogames on modern home PCs. The first version of MAME was released by an Italian developer in 1997. Users of this free software download old arcade videogames (zipped in one file named "rom") from the web. Many of them are abandonware[6] and are freely playable by users without any copyright issue. MAME runs roms and executes them as normal PC videogames playable via keyboard.

The success of this vintage innovation is based on two different communities. The first is the restricted community of developers sharing a strong passion for old arcade videogames and working directly on the user innovation process. From the point of view of programmers, MAME is an effort to develop and improve hacking practices and capabilities (Murphy 2013). Developers contribute actively and without remuneration to the software update and increase the amount of games freely downloadable by end-users, which constitute the second "virtual" community related to MAME. MAMEWorld is one of the most influential virtual communities focused on MAME. This distinction between users-innovators and users-consumers is typical of many online communities developing open source software (Von Hippel 2003).

Figure 5.2 reports the evolution of ROMs development for MAME32 between 1997 and 2008. The blue chart reports the number of ROMs developed by programmers. The red chart reports the number of unique videogames available to users (more ROMs could be developed for the same videogame).

MAME offers an experience simulated and, thus, totally different from the original emotion of playing an arcade videogame in a game center. However, users can recover and exploit another time their expertise, achieved with much effort and many years ago. MAME allows multiplayer games up to eight players and provides many playing options (e.g., controls) for customizing the game experience and practice of nostalgic players. Emulators convert arcade videogames in common home videogames that users can play without any time limitation

---

[6] Abandonware is old and abandoned software (in this case, a videogame) for which the producer does not claim anymore property rights.

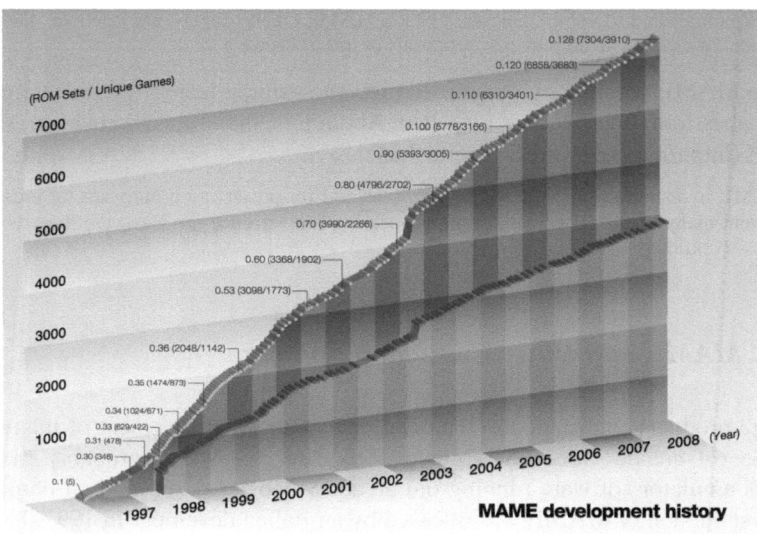

**Fig. 5.2**  The evolution of the MAME project (source: www.mamdedev.org)

(for instance, due to the limited opening hours of videogame centers) and expenditure of money per game (from here comes the famous sentence "Insert Coin" appearing on the screen at the end of each game).

MAME is not the only emulation software of old videogames. There are many other cases of emulators establishing interoperability between old arcade videogames and modern PCs with hardware designed for new generation videogames. Other emulators reproduce old videogames originally designed for home consoles or PC (e.g., WinUAE for Amiga 500 videogames). In all these cases, emulators are not just reactions to preserve declining knowledge, videogames, and practice by an existing community of aficionados but also drivers of further development of the communities themselves. The members of these communities manage thematic websites in which they share information and suggestions about old arcade videogames and their emulators, additional utilities, and hardware (e.g., joysticks).

Table 5.2 summarizes the main emulators for old arcade videogames currently available on Internet.

Emulators clearly modify the traditional experiences of users and aesthetics of arcade videogames. However, the improvement in performance of old arcade videogames by MAME and other emulators is evident and, at least, twofold:

- Emulators improve the outputs of the complementary service "inter-operability with external devices." The fruition of old arcade videogames was constrained by the technical characteristics of the cabinets (e.g., number of joysticks, monitor size, sound card) in which they were embedded. MAME allows videogamers to extend somehow and customize the set of external devices by which playing

**Table 5.2**  Main emulators for old arcade videogames

| Emulator | Type | Platform | Original system emulated |
|---|---|---|---|
| Mame32 | Freeware | Windows | Multiple systems (mainly videogames in game centers) |
| Nebula | Freeware | Windows | Neo Geo, CPS1/2, Konami and PGM (PolyGameMaster) |
| Kawaks | Freeware | Windows | Neo-Geo and Capcom (CPS1/2) |
| Zinc | Freeware | Windows | Sony ZN1, ZN2, and Namco System 11 |
| Calice | Freeware | Windows | Capcom CPS-1, Capcom CPS-2, Sega System 16, Sega System 18, SNK Neogeo, and Gaelco System 1 |

games (e.g., by using also PC keyboard or decreasing the audio level). In other words, emulators transform home PCs in highly personalized cabinets.

• Emulators improve the outputs of the complementary service "human-software interaction." As above, this vintage product allows users to customize greatly their interaction with the arcade videogame by improving dramatically the number of set-up options available for players. For instance, video-gamers can set ex-ante (before launching the MAME rom) their preferences about user interface and controls during the game, selections rarely possible with the traditional videogames versions. Therefore, users now send more commands to software.

These improvements increase the effectiveness for customers of traditional videogame players. The core of these improvements lies in the augmented personalization in the fruition of the videogame. However, the greatest benefit of this vintage product for customer is the possibility to play these arcade videogames still nowadays, after many years from their "commercial death," and live again the emotions and feelings felt when these aficionados were kids or teenagers.

## 5.4  Conclusions and Implications

Also, customers can generate autonomously vintage innovation in order to improve the OTBP's performance and their own effectiveness. A very interesting evidence is that this form of vintage innovation can be developed by both existing and established old technology-based communities (radio-amateurs) and emerging communities, formed almost from scratch in order to gain not just more performance and effectiveness but also a "life after death" for their old beloved technological products in decline (MAME developers and users).

Both the cases show user innovation in these old technology-based communities was basically a reaction to technological change aimed at improving and/or preserving their declining practices and activities. However, a critical difference between MAME users and radio-amateurs talking by Echolink or IRLP is the extent of the support of each community to the vintage product. Aficionados of old arcade videogames look more oriented, overall, than radio-amateurs to support the diffusion of the vintage product in their community. A number of motivations could explain this difference. First, the greater establishment and history of ham radios makes some purist users more skeptical about interoperability with new technology. Second, the vintage product for videogames players is necessary for the use of their OTBP and not optional as for radio-amateurs. Third, the issue of preservation and risk of disappearing are much more critical for arcade videogames than ham radios.

The managerial implications of this autonomous form of vintage innovation are closely related to the issue of exploiting community ambivalence for new technology and the outputs of their user innovation process. First, managers or key members of old technology-based CoPs (e.g., professional communities) should plan regularly activities of technology stewardships for community members. The critical issue for companies in order to maximize the innovative potential of these communities is to identify the innovators within these communities, support, and exploit them to create value. In general, community members are likely to be oriented and interested to innovation (experimentation, as in the radio-amateurs case) (Brown and Duguid 1991). An effective selection by firms of proper innovators and change agents is critical in order to support this process. Some specialized members of the community will develop the vintage product making interoperable the OTBP and new technology. In addition, key members of old technology-based CoPs should orient properly the community toward positive perceptions about vintage products.

Companies should try to govern and/or support somehow these processes of user innovation and diffusion of innovation in order to exploit better the opportunities deriving from the ambivalent reactions of vintage communities to technological change. This logic is basically different from the standard view of co-development of innovation based on the "Customer Active Paradigm" (Von Hippel 2003). These cases show communities of old technology users as the real actors promoting innovation. Companies should just follow, support, and take advantage of their innovation-related efforts.

Emulation can be a very suitable means for the improvement of performance of OTPS and customer effectiveness when vintage innovation is developed by users. Indeed, this is likely to be the most suitable and easiest mode for communities to establish backwards compatibility. This view implies that many communities of users probably could be facilitated in preserving and improving the performance of their traditional practices (communicating by VOIP and gaming) rather than their outdated products (ham radios and arcade videogames). This consideration is obvious when emulation remains the only way for the survival and fruition of the OTBP.

Referring to the domain of technology management, the most common situation for vintage innovation by users is that community members establish backwards compatibility between their OTBP and new technology. The likelihood of this process is made easier by the rise of information technology supporting emulation and, overall, inter-operability between different platforms (Farrell and Saloner 1992). The cases show this form of vintage innovation occurred for OTBPs embedded within quite large technological systems. Conversely, one might expect that users should not have enough technical equipment, knowledge, or resources to upgrade the performance of OTBPs of complex and large systems (as turntables or analog cameras). Instead, these evidences outline that users can innovate even within complex modular systems by exploiting their competences in various technological fields (as informatics or electronics). Companies, thus, could support vintage innovation by users also for further types of more networked and complex OTBPs.

Referring to innovation strategy of technological organizations, this form of vintage innovation provides several areas of actions for companies. For instance, companies should support creative and innovative efforts of communities for vintage innovation by providing specific toolkits. Users will use these tools for developing innovations improving their traditional technological practices and products becoming obsolete.

Finally, this type of vintage innovation has various implications in the domain of marketing. Every marketing action should be targeted to those members of the community playing the roles of innovators and/or change agents. These individuals will play the functions of community lead users (Von Hippel 2003) promoting the entry of new technology within their social aggregation and, thus, companies should be able to identify them and understand which incentives might support (and which costs might interfere) their innovation-related efforts.

# References

Amis M (1982) Invasion of the space invaders. Hutchinson, London

Bogdan C, Bowers J (2007) Tuning in: challenging design for communities through a field study of radio amateurs. In: Proceedings of the communities and technologies 2007, Springer, London, pp 439–461

Bogers M, Afuah A, Bastian B (2010) Users as innovators: a review, critique, and future research directions. J Manage 36(4):857–875

Brown JS, Duguid P (1991) Organizational learning and communities-of-practice: toward a unified view of working, learning, and innovation. Organ Sci 2(1):40–57

Corts KS, Lederman M (2009) Software exclusivity and the scope of indirect network effects in the US home video game market. Int J ind Organ 27(2):121–136

Farrell J, Saloner G (1992) Converters, compatibility, and the control of interfaces. J Ind Econ 40:9–36

Faulkner P, Runde J (2009) On the identity of technological objects and user innovations in function. Acad Manag Rev 34(3):442–462

Fernández-Vara C (2009) Play's the thing: a framework to study videogames as performance. In: Proceedings of DiGRA, innovation in games, play, practice and theory

Gernsback H (1909) First annual official wireless blue book of the wireless association of America. Modern Electrics Publication, New York

Hague P, Hague N, Morgan CA (2004) Market research in practice: a guide to the basics. Kogan Page, London

Haring K (2007) Ham Radio's technical culture. MIT Press, Cambridge

Herman L, Horwitz J, Kent S, Miller S (2002) The history of video games. Gamespot. Retrieved
    February, 7, 2002
Murphy D (2013) Hacking public memory understanding the multiple arcade machine emulator.
    Games Cult 8(1):43–53
Orlikowski WJ (2000) Using technology and constituting structures: a practice lens for studying
    technology in organizations. Organ Sci 11(4):404–428
Poole S (2004) Trigger happy: videogames and the entertainment revolution. Arcade Pub, New
    York
Scapens RW (2004) Doing case study research. In: Humphrey C, Lee B (eds) The real life guide
    to accounting research. Elsevier, New York, pp 257–279
Schiavone F (2011) User innovation e cambiamento tecnologico nelle tribù hightech. Mercati e
    Competitività 2011/4(4):167–184
Schiavone F (2012) Resistance to industry technological change in communities of practice: the
    "ambivalent" case of radio amateurs. J Organ Change Manage 25(6):784–797
Schiavone F, Agrifoglio R (2012) Communities of practice and practice preservation: a case
    study. In: De Marco M, Te'eni D, Albano V, Za S (eds) Information systems: crossroads for
    organization, management, accounting and engineering. Physica-Verlag HD, New York, pp
    331–338
Valdes R, Roos D (2001) "How VoIP works," available at: http://computer.howstuffworks.com/
    ip-telephony.htm (accessed 14 August 2011)
Yin R (1994) Case study research: design and methods. Sage, Beverly Hills
Von Hippel E (2001) Learning from open-source software. MIT Sloan manag rev 42(4):82–86
Von Hippel E (2003) Democratizing innovation. MIT Press Books, Boston